Lecture Notes in Networks and Systems 1769

The series "Lecture Notes in Networks and Systems" publishes the latest developments in Networks and Systems—quickly, informally and with high quality. Original research reported in proceedings and post-proceedings represents the core of LNNS.

Volumes published in LNNS embrace all aspects and subfields of, as well as new challenges in, Networks and Systems.

The series contains proceedings and edited volumes in systems and networks, spanning the areas of Cyber-Physical Systems, Autonomous Systems, Sensor Networks, Control Systems, Energy Systems, Automotive Systems, Biological Systems, Vehicular Networking and Connected Vehicles, Aerospace Systems, Automation, Manufacturing, Smart Grids, Nonlinear Systems, Power Systems, Robotics, Social Systems, Economic Systems and other. Of particular value to both the contributors and the readership are the short publication timeframe and the world-wide distribution and exposure which enable both a wide and rapid dissemination of research output.

The series covers the theory, applications, and perspectives on the state of the art and future developments relevant to systems and networks, decision making, control, complex processes and related areas, as embedded in the fields of interdisciplinary and applied sciences, engineering, computer science, physics, economics, social, and life sciences, as well as the paradigms and methodologies behind them.

Indexed by SCOPUS, EI Compendex, INSPEC, WTI Frankfurt eG, zbMATH, SCImago.

All books published in the series are submitted for consideration in Web of Science.

For proposals from Asia please contact Aninda Bose (aninda.bose@springer.com).

Elżbieta Macioszek · Irena Jurdana ·
Grzegorz Sierpiński
Editors

Smart and Environmentally Friendly Contemporary Road Traffic and Transportation Engineering Solutions

Springer

Editors
Elżbieta Macioszek
Silesian University of Technology
Gliwice, Poland

Irena Jurdana
University of Rijeka
Rijeka, Croatia

Grzegorz Sierpiński
Silesian University of Technology
Gliwice, Poland

ISSN 2367-3370 ISSN 2367-3389 (electronic)
Lecture Notes in Networks and Systems
ISBN 978-3-032-14093-7 ISBN 978-3-032-14094-4 (eBook)
https://doi.org/10.1007/978-3-032-14094-4

This Springer imprint is published by the registered company Springer Nature Switzerland AG
The registered company address is: Gewerbestrasse 11, 6330 Cham, Switzerland

Preface

Transport and its organization are constantly changing. The constantly increasing traffic intensity causes a decrease in the attractiveness of cities and regions, which in turn forces organizational, structural, and legal changes related to road traffic. This is related to the progressive expansion of urban space, as well as the development of technology used in this type of transport. Against this background, smart, and environmentally friendly contemporary road traffic and transportation engineering solutions are vital for the advancement of sustainable transportation and promotion of efficient urban mobility.

The presented volume entitled "*Smart, and Environmentally Friendly Contemporary Road Traffic and Transportation Engineering Solutions*" provides a good opportunity to learn about the consequences of increasing traffic volumes along with the advanced solutions used to improve the efficiency of functioning existing transport networks by implementing new innovative methods of traffic control and traffic management. This knowledge was acquired after the conducted road traffic research, analysis, and modeling in transport networks. Some of the topics included are characterized by a practical approach to solving current problems in traffic engineering and road transport. The book has been divided into three parts. There are:

- Part 1. Road Traffic Engineering, and Road Traffic Safety.
- Part 2. Smart, and Ecological Solutions.
- Part 3. Economic, Regulatory, and Social Aspects.

The content of the volume consists of selected papers submitted and presented at the 21th Scientific and Technical Conference "Transport Systems. Theory and Practice", organized by the Department of Transport Systems, Traffic Engineering and Logistics at the Faculty of Transport and Aviation Engineering of the Silesian University of Technology (Katowice, Poland). Each article has been reviewed by two independent Reviewers (the so-called double blind review). Among the topics discussed were issues related to current problems in the field of the theoretical approaches and practical solutions for road traffic and road transportation engineering solutions. Numerous practical examples illustrate the possibilities of improving the current situation with a view to human welfare, sustainable development of transport systems, and environmental protection.

Taking the opportunity, we would like to thank the Authors for the submitted papers, for their great contribution to illustrating the numerous advanced solutions and practical applications in road traffic research, analysis, and modeling, as well as for sharing the results of their scientific and research works. We also thank the Reviewers for their insightful comments and suggestions that helped keep the volume of the book high-quality.

We also wish the Readers a successful reading.

October 2025

Elżbieta Macioszek
Irena Jurdana
Grzegorz Sierpiński

Organization

21th Scientific and Technical Conference "Transport Systems. Theory and Practice" (TSTP2025) is organized by the Department of Transport Systems, Traffic Engineering and Logistics, Faculty of Transport and Aviation Engineering, Silesian University of Technology (Poland).

Organizing Committee

Organizing Chair

Grzegorz Sierpiński	Silesian University of Technology, Poland

Members

Agnieszka Popowicz
Ireneusz Celiński
Adrian Barchański
Iga Przytuła
Marcin Kłos
Wioletta Cebulska
Krystian Szewczyński
Piotr Soczówka
Sofia Nardoianni
Maria Luisa Tumminello

The Conference took place under the Honorary Patronage

Ministry of Infrastructure
Voivode of Silesia
Marshal of the Silesian Voivodeship

Partners of the Conference

European Platform of Transport Sciences

National Road Safety Centre (in Polish: Krajowe Centrum Bezpieczeństwa Ruchu Drogowego)

Centre for European Union Transport Projects (in Polish: Centrum Unijnych Projektów Transportowych)

Scientific Committee

Grzegorz Sierpiński (Co-chair)	Silesian University of Technology, Poland
Elżbieta Macioszek (Co-chair)	Silesian University of Technology, Poland
Rahmi Akçelik	SIDRA SOLUTIONS, Australia
Monika Bąk	University of Gdańsk, Poland
Werner Brilon	Ruhr-University Bochum, Germany
Jacek Chmielewski	Cracow University of Technology, Poland
Margarida Coelho	University of Aveiro, Portugal
Csaba Csiszár	Budapest University of Technology and Economics, Hungary
Mauro D'Apuzzo	University of Cassino and Southern Lazio, Italy
Roberta Di Pace	University of Salerno, Fisciano, Italy
Mehmet Dikmen	Baskent University, Turkey
Marek Drličiak	University of Žilina, Slovakia
Grzegorz Dydkowski	University of Economics in Katowice, Poland
Domokos Esztergár-Kiss	Budapest University of Technology and Economics, Hungary
Zoltán Fazekas	Institute for Computer Science and Control, Hungary
Piotr Folęga	Silesian University of Technology, Poland
József Gál	University of Szeged, Hungary
Anna Granà	University of Palermo, Italy
Andrzej S. Grzelakowski	Gdynia Maritime University, Poland
Mehmet Serdar Güzel	Ankara University, Turkey
Józef Hansel	AGH University of Science and Technology, Cracow, Poland
Libor Ižvolt	University of Zilina, Slovakia
Marianna Jacyna	Warsaw University of Technology, Poland
Ilona Jacyna-Gołda	Warsaw University of Technology, Poland
Kazimierz Jamroz	Gdańsk University of Technology, Poland
Nan Kang	Tokyo University of Science, Japan
Grzegorz Karoń	Silesian University of Technology, Poland
Jan Kempa	Bydgoszcz University of Science and Technology, Poland
Michael Koniordos	Pireaus University of Applied Sciences, Greece
Stanisław Krawiec	Silesian University of Technology, Poland

Bogusław Łazarz	Silesian University of Technology, Poland
Zbigniew Łukasik	Kazimierz Pulaski University of Technology and Humanitiesin Radom, Poland
Michal Maciejewski	Technical University Berlin, Germany
Ján Mandula	Technical University of Košice, Slovakia
Antonio Masegosa	IKERBASQUE Research Fellow at University of Deusto Bilbao, Spain
Agnieszka Merkisz-Guranowska	Poznań University of Technology, Poland
Anna Mężyk	Kazimierz Pulaski University of Technology and Humanities in Radom, Poland
Leszek Mindur	University College of Professional Education in Wroclaw, Poland
Maciej Mindur	Lublin University of Technology, Poland
Goran Mladenović	University of Belgrade, Serbia
Kai Nagel	Technical University Berlin, Germany
Piotr Niedzielski	University of Szczecin, Poland
Przemysław Niewiadomski	University of Zielona Góra, Poland
Piotr Olszewski	Warsaw University of Technology, Poland
Enrique Onieva	Deusto Institute of Technology University of Deusto Bilbao, Spain
Asier Perallos	Deusto Institute of Technology University of Deusto Bilbao, Spain
Hrvoje Pilko	P3Mobility Ltd, Zagreb, Croatia
Antonio Pratelli	University of Pisa, Italy
Dariusz Pyza	Warsaw University of Technology, Poland
Cesar Queiroz	World Bank Consultant (Former World Bank Highways Adviser), Washington, DC, USA
František Schlosser	University of Žilina, Slovakia
Ewelina Sendek-Matysiak	Kielce University of Technology, Poland
Jacek Skorupski	Warsaw University of Technology, Poland
Aleksander Sładkowski	Silesian University of Technology, Poland
Aleksander Sobota	Silesian University of Technology, Poland
Marcin Staniek	Silesian University of Technology, Poland
Wiesław Starowicz	Cracow University of Technology, Poland
Andrzej Szarata	Cracow University of Technology, Poland
Antoni Szydło:	Wroclaw University of Technology, Poland
Grzegorz Ślaski	Poznań University of Technology, Poland
Marko Šoštarić	University of Zagreb, Croatia
Andrew P. Tarko	Purdue University, West Lafayette, USA
Dariusz Tłoczyński	University of Gdańsk, Poland
Agnieszka Tubis	Wroclaw University of Technology, Poland
Frane Urem	Polytechnic of Šibenik, Croatia

Hua-lan Wang	School of Traffic and Transportation, Lanzhou Jiaotong University, Lanzhou, China
Mariusz Wasiak	Warsaw University of Technology, Poland
Adam Weintrit	Gdynia Maritime University, Poland
Andrzej Więckowski	AGH University of Science and Technology, Cracow, Poland
Adam Wolski	Polish Naval Academy, Gdynia, Poland
Marcin Wołek	University of Gdańsk, Poland
Elżbieta Załoga	University of Szczecin, Poland
Jacek Żak	Poznań University of Technology, Poland
Jolanta Żak	Warsaw University of Technology, Poland
Renata Żochowska	Silesian University of Technology, Poland

Referees

Piotr Czech
Elżbieta Macioszek
Grzegorz Wojnar
Grzegorz Karoń
Grzegorz Sierpiński
Renata Żochowska
Marcin J. Kłos
Marcin Staniek

Contents

Road Traffic Engineering, and Road Traffic Safety

Analyzing the Relationship Between Built Road Environment and Perceived Road Safety. A Case Study in Four Cities in Flanders, Belgium 3
Girma Fikre, An Neven, and Davy Janssens

Setting Priorities for Safety Improvements on Two-Lane Roads in Israel 16
Victoria Gitelman

Not Just the Driver: How Road Design Influences Speeding and Traffic Safety 30
Marko Šoštarić, Marijan Jakovljević, Juraj Leonard Vertlberg, and David Gruhonjić

Hourly Distribution of the Probability of a Fire Accident in a Urban Road Tunnel 42
Aleksander Król, Małgorzata Król, and Krystian Szewczyński

Evaluation of the Effectiveness of Traffic Calming Measures on the City's Street and Road Network 55
Olha Kholodova, Maryna Buhaiova, Andrii Sakhno, and Oleh Kholodov

Analyzing and Evaluating Safety in Numbers for Cyclists in a Large Dataset ... 75
Andreas Leich, Ronald Nippold, Louis Calvin Touko Tcheumadjeu, and Peter Wagner

Smart, and Ecological Solutions

A Microsimulation-Driven Approach to Assess Performance of Alternative Intersection Designs Embedded with Smart Vehicle Technologies 89
Anna Granà, Elżbieta Macioszek, and Maria Luisa Tumminello

Towards Accurate Vehicle Speed Estimation in Aerial Views: How Localization and Tracking Methods Influence Speed Estimation in UAV-Based Vehicle Monitoring 106
Muhammad Waqas Ahmed, Muhammad Adnan, Muhammad Ahmed, Davy Janssens, Geert Wets, Afzal Ahmed, and Wim Ectors

A Virtual Approach to Evaluate Vehicular Flows to Promote Sustainable and Safe Mobility in Small and Smart Communities 122
Giuseppe Cappelli, Sofia Nardoianni, Mauro D'Apuzzo, and Vittorio Nicolosi

The Impact of Road Conditions on Energy Consumption in the Electric Drive System of a Passenger Car – Experimental Analysis in the RDE Cycle ... 138
Maciej Sproch, Mariusz Graba, Krystian Hennek, and Jarosław Mamala

Leveraging GAIA-X for Intelligent Mobility and Transportation Systems in Europe 154
Sebastian Naumann and Joachim Schade

Digital Twin of a Specialized Off-Road Vehicle for Peoples with Special Needs 170
Marcin Staniek and Ireneusz Celiński

Economic, Regulatory, and Social Aspects

Involvement of Pension Funds in Public Private Partnership Projects 197
Václav Baroch and Rudolf Franz Heidu

Factors Shaping the Supply of Traditional Fuels and Their Impact on Price, Analysis of Selected Cases 210
Piotr Kotylak

Preliminary Diagnosis of the Impact Factors of the COVID-19 Pandemic on Passenger Rail Transport Based on a Short Survey 225
Katarzyna Markowska

Author Index 245

Road Traffic Engineering, and Road Traffic Safety

Analyzing the Relationship Between Built Road Environment and Perceived Road Safety. A Case Study in Four Cities in Flanders, Belgium

Girma Fikre[1,2](✉), An Neven[1], and Davy Janssens[1]

[1] Transportation Research Institute (IMOB), Hasselt University, Martelarenlaan 42, 3500 Hasselt, Belgium
{girma.fikre,an.neven,davy.janssens}@uhasselt.be

[2] Department of Civil Engineering, Jimma University, Jimma Institute of Technology, Jimma, Ethiopia

Abstract. Understanding the impact of urban infrastructure on perceived road safety is crucial for improving children's road safety. This study explores the relationship between the built environment and perceived safety using a novel combination of subjective perception data and geospatial data. Perceived safety assessments were gathered through the Route2School (R2S) application, which collects crowdsourced inputs from schoolchildren and their guardians. These inputs were linked with detailed road infrastructure data from OpenStreetMap (OSM) across four cities in Flanders, Belgium: Antwerp, Gent, Mol, and Zonhoven.

The study's findings reveal both universal and city-specific determinants of perceived safety. While road type, presence and type of intersection, and bicycle infrastructure emerged as common influential features, their importance varied across the urban contexts.

The approach presented here demonstrates how integrating crowdsourced data with open-source spatial information can support child-centric urban design and targeted policy actions to enhance perceived traffic safety and mobility. The insights gained from this study can inform data-driven planning and policy-making aimed at creating safer, more inclusive urban environments for vulnerable road users.

Keywords: Perceived road safety · Built Road · Crowdsourced data

1 Introduction

Children's mobility and their perception of safety in urban environments are increasingly recognized as important indicators of inclusive and sustainable city planning [1]. The degree to which children feel safe in public spaces significantly influences their willingness to walk or cycle to school, participate in outdoor activities, and explore their neighborhoods [2]. Despite improvements in physical infrastructure and traffic management, many children and their caregivers continue to perceive urban streets as unsafe [3].

E. Macioszek et al. (Eds.): TSTP 2025, LNNS 1769, pp. 3–15, 2026.
https://doi.org/10.1007/978-3-032-14094-4_1

Traditional traffic safety analyses focus on crash statistics and physical design standards [4]. However, such objective metrics may not fully capture the experiences and fears of child pedestrians and cyclists. For example, a road segment with low accident frequency may still feel unsafe due to high vehicle speeds, lack of visibility, or poorly maintained pedestrian infrastructure. There is growing recognition of the need to supplement conventional safety evaluations with subjective measures of perceived safety, especially from vulnerable road users such as children [5].

In recent years, crowdsourced applications have become valuable tools for collecting data for transportation studies [6]. Crowdsourced applications like the Route2School (R2S) [7] application empower users to report real-time perceptions of safety, allowing researchers and policymakers to access rich, localized data on user experience. The Route2School (R2S) platform has been used to collect user-generated reports identifying traffic bottlenecks along children's school routes in Belgium. Each report is georeferenced and includes information on the type of problem, a description, the mode of transport involved, and suggested solutions. When combined with detailed spatial data such as that from OpenStreetMap (OSM) [8], these user insights can offer a powerful foundation for understanding how specific features of the road environment contribute to feelings of safety or discomfort.

Several studies have examined the built road-related factors influencing children's perceptions of road safety. A study by [9] shows that the school neighborhood-built environment significantly correlates with adolescent road safety perception. Prior research shows that road width, crosswalks, student-friendly facilities at the intersection, and four-way intersections were significant and positively associated with perceived crash risk among school-aged children [10]. Another study related children's road safety perception to walking, fast traffic, people's behavior, and small cities [11].

Despite the significant body of research, there remain several gaps in understanding the nuanced relationship between road infrastructure and the perceived safety of children. While there are numerous studies on infrastructure features, few have examined how children's personal experiences or crowd-sourced data (e.g., from apps like Route to School) can inform the broader understanding of safety perceptions. Moreover, while existing studies often focus on individual infrastructural elements, fewer have examined how the cumulative effect of multiple infrastructure features interacts with perceived safety.

Additionally, many studies have focused on general urban settings, with limited research on how these dynamics apply to specific regions, such as big and small cities. Understanding these differences is crucial, as infrastructure-related safety concerns may vary significantly between big and small towns.

This paper contributes to filling these research gaps by integrating data from the R2S app and OSM to study the relationship between road infrastructure and the perceived road safety of children in four cities in Flanders, Belgium. By leveraging both objective data from the infrastructure and subjective data from crowdsourced reports, this research provides a comprehensive understanding of how the built environment shapes children's safety perceptions.

This paper investigates the relationship between built road environment characteristics and perceived road safety among schoolchildren in four Flemish cities. We integrate geospatial data with crowdsourced perceptions and apply Random Forest (RF) machine learning models to identify the most critical features influencing perceived safety. Furthermore, we employ Shapely Additive Explanation (SHAP) values to interpret each variable's role, providing predictive and explanatory insights.

The remainder of this paper is structured as follows. Section 2 provides a detailed overview of the methodological approach, including the data sources, preprocessing steps, and the methods used in the analysis. Section 3 presents the results, focusing on the relationship between the built road environment and perceived risk in four cities. Section 4 discusses the findings in relation to existing literature and highlights their practical implications. Finally, Sect. 5 concludes the paper by summarizing the key findings and offering suggestions for future research.

2 Methodology

This study integrates multiple data sources and advanced analytical techniques to investigate the relationship between built road environment characteristics and perceived road safety. Four main steps were followed in this study, as illustrated in the chart below. These steps outline the overall research workflow, from data collection to model interpretation (FIg. 1).

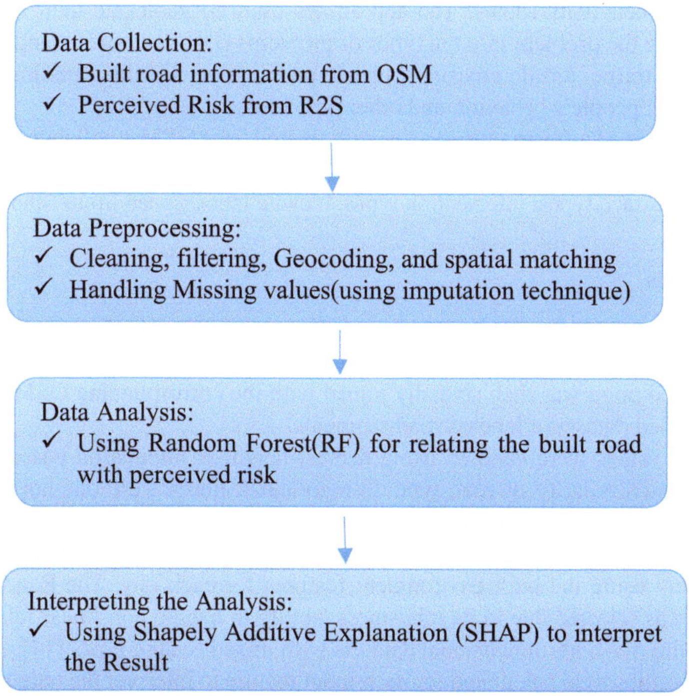

Fig. 1. Chart Showing the steps followed in the research

2.1 Study Area and Data Collection

The research focuses on four Flemish cities: two large cities, such as Antwerp and Gent, and two smaller cities, such as Mol and Zonhoven. These cities were selected due to their participation in the Route2School (R2S) program, the availability of both perception data and geospatial datasets and the distinctive differences between the cities. Table 1 shows a summary of the cities.

Table 1. Cities Considered for the Study [7, 12, 13]

Municipality	Reported Bottleneck	Population in January 2024	Total of accidents per 100 km of road in 2022
Antwerp	1021	544,759	98.79
Gent	1187	269,597	56.77
Mol	282	38,534	15.11
Zonhoven	307	21,845	20.2

Perception data were collected through the R2S application, which enables schoolchildren and their guardians to report safety perceptions of specific road segments along their daily routes. The app allows users to geolocate their observations and categorize the problem into ten types of problems (i.e., bad roads, dangerous intersections, fast traffic, unsafe crossings, unsafe paths, poor visibility, missing crossings, missing paths, people's behavior, and other problems).

In parallel, road infrastructure data were extracted from OSM, a collaborative geospatial database that provides detailed information on different road attributes such as road types, road surface types, intersection types, bicycle lanes, speed limits, and land use.

2.2 Analysis

The collected perception data were preprocessed to remove duplicates and irrelevant entries. Only reports linked to specific and identifiable road segments were retained. Each reported point was then spatially joined with the corresponding OSM features to create a unified dataset of labeled road segments.

Missing values were handled using nearest-neighbor imputation based on spatial proximity and similarity of road type. Categorical features were one-hot encoded to facilitate use in machine learning models.

For each city, a separate Random Forest (RF) classifier was trained to predict perceived safety using the built environment features for each city. The Random Forest algorithm was selected due to its robustness, ability to handle non-linear relationships, and suitability for high-dimensional datasets with mixed feature types [14].

SHAP values were calculated for each input feature to interpret the output of the RF models. SHAP is a model-agnostic explanation technique based on cooperative game theory, which attributes the contribution of each feature to the model's prediction for individual instances [15].

We used TreeSHAP, an efficient algorithm designed specifically for tree-based models, to compute global and local feature importances. Global SHAP summaries reveal the overall ranking and direction of influence of features across the entire dataset, while local SHAP plots illustrate how features contribute to individual predictions.

SHAP analysis provided critical insights into which infrastructure elements consistently contribute to the type of perceived safety and how these effects vary across different urban contexts.

3 Results

This section presents the machine learning models and SHAP-based interpretation findings, focusing on identifying the most influential built environment features shaping perceived road safety in each of the four cities studied: Antwerp, Gent, Mol, and Zonhoven. Figure 2 shows a SHAP summary plot of the four cities in perceiving dangerous intersections. In the SHAP summary plot, each point represents a single observation (i.e., a reported bottleneck), with the color indicating the value of the feature—red for high values and blue for low values. The variables are ranked by their overall importance, with the most influential features appearing at the top of the figure. These top-ranked features significantly impact predicting whether a location is perceived as a dangerous intersection.

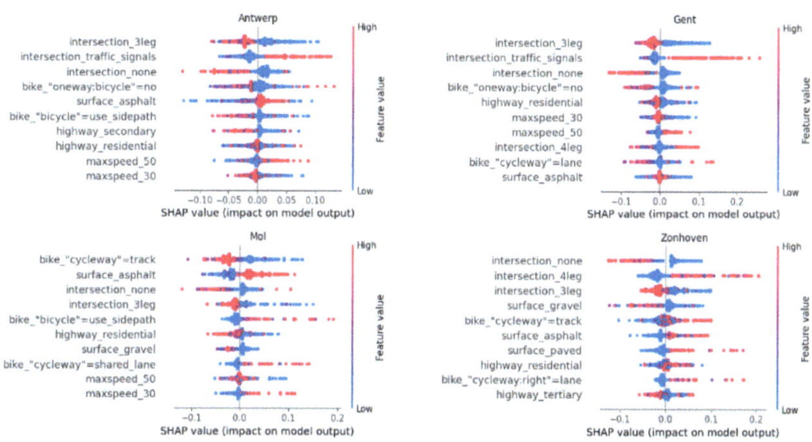

Fig. 2. SHAP summary plot showing the relation between road features and perceiving a dangerous intersection

In contrast, variables positioned toward the bottom of the plot have a weaker association with the perceived safety issue. Whenever red values are predominant on the positive side, it implies the feature has a positive effect. For instance, among the many features of Antwerp and Gent cities, traffic signals have the highest positive impact on perceiving dangerous intersections. Meanwhile, in Mol City, asphalt roads have the highest positive impact on perceiving dangerous intersections, whereas, in Zonhoven, four-leg intersections have the highest positive impact on perceiving dangerous intersections. Similarly, SHAP summary plots were generated for all ten perceived safety problem types reported in the R2S application across the four selected cities. Based on these visualizations, Tables 2, 3, 4 and 5 present the most influential variable, i.e., the variable with the highest positive SHAP impact identified for each problem type in each city.

Table 2. Average SHAP values of road features which has the highest positive effect on perceived risk in Antwerp

	Road Type	Road Surface	Intersections	Land Use	Maximum Speed	Bike lanes
Bad Road	Tertiary (0.008)	Cobblestones (0.015)	Three leg (0.015)	Meadow (0.004)	30 (0.006)	Cyclestreet (0.018)
Dangerous Intersection	Primary (0.003)	-	Traffic signals (0.022)	-	50 (0.007)	Bike Lane track at right(0.005)
Fast Traffic	Residential (0.016)	Asphalt (0.014)	Midblock (0.043)	Forest (0.002)	30 (0.007)	Cyclestreet (0.014)
Missing Crossing	Residential (0.01)	Concrete (0.005)	Three leg (0.018)	Park (0.002)	-	bidirectional bikelane (0.018)
Missing Path	-	-	Midblock (0.016)	-	-	Bike allowed on the road (0.007)
People Behavior	Secondary (0.01)	Cobblestones (0.008)	Three leg (0.013)	Residential (0.003)	30 (0.007)	bidirectional bike lanes (0.012)
Poor Visibility	Residential (0.013)	Asphalt (0.015)	Three leg (0.024)	Residential (0.003)	-	bidirectional bikelane (0.032)
Unsafe Crossing	Tertiary (0.006)	paved (0.002)	Four leg (0.013)	-	70 (0.003)	Use sidepath (bike not allowed) (0.01)
Unsafe Path	Secondary (0.008)	paving stones (0.005)	-	Forest (0.002)	-	one-way bike lane (0.007)
Other	Tertiary (0.012)	-	Traffic Signals (0.013)	Park (0.004)	-	bidirectional bike lanes (0.008)

Table 3. Average SHAP values of road features which has the highest positive effect on perceived risk in Gent

	Road Type	Road Surface	Intersections	Land Use	Maximum Speed	Bike lanes
Bad Road	Residential (0.016)	Cobblestones (0.006)	Round-about (0.017)	Residential (0.004)	30 (0.015)	Bike allowed (0.028)
Dangerous Intersection	-	-	Traffic signals (0.023)	Scrub (0.002)	50 (0.008)	Bike allowed on the road (0.006)
Fast Traffic	Residential (0.008)	Asphalt (0.008)	Midblock (0.011)	Forest (0.001)	30 (0.009)	bike lanes provided (0.018)
Missing Crossing	Residential (0.014)	gravel (0.015)	Three leg (0.016)	Residential (0.009)	50 (0.008)	bike lane track at the right (0.024)
Missing Path	Residential (0.014)	gravel (0.018)	Midblock (0.017)	Scrub (0.002)	50 (0.006)	bidirectional bike lane (0.028)
People Behavior	Tertiary (0.014)	Paved (0.002)	-	-	30 (0.018)	no bikelanes (0.007)
Poor Visibility	Residential (0.013)	-	Midblock (0.011)	Park (0.002)	-	bidirectional bikelane (0.015)
Unsafe Crossing	Primary (0.005)	paving stones (0.008)	Traffic Signals (0.013)	Park (0.003)	-	-
Unsafe Path	Tertiary (0.008)	Concrete (0.003)	Mid-block (0.025)	Park (0.005)	50 (0.007)	bike lane at the right (0.007)
Other	Trunk (0.002)	–	-	-	-	-

The model indicates that residential roads are associated with reports of fast traffic, missing crossings, and poor visibility across three of the four cities. In Ghent and Zonhoven, children are likelier to report missing paths on residential roads. Additionally, reports of bad roads are widespread in Antwerp and Zonhoven. Tertiary roads are more commonly associated with complaints about people's behavior, especially in Gent and Zonhoven. Unsafe paths are notably reported on residential roads in Gent and Mol. Regarding primary roads, children in Antwerp and Mol are more likely to report dangerous intersections, while unsafe crossings are more commonly reported in Gent and Mol.

The results show that intersections play a significant role in shaping children's perceptions of road safety. The model reveals that traffic signals and four-leg intersections are often perceived as dangerous and unsafe for crossing. Missing crossings are most frequently reported at three-leg and four-leg intersections, highlighting the importance of clear and accessible pedestrian infrastructure at these locations. Moreover, fast traffic and missing paths are predominantly reported along midblock road segments far from intersections. The model further shows that in Ghent and Mol, midblock segments are strongly associated with perceptions of unsafe paths.

Table 4. Average SHAP values of road features which has the highest positive effect on perceived risk in Mol

	Road Type	Road Surface	Intersections	Land Use	Maximum Speed	Bike lanes
Bad Road	Primary (0.015)	ground (0.01)	Midblock (0.023)	Meadow (0.008)	30 (0.02)	bidirectional bike lanes (0.012)
Dangerous Intersection	Primary (0.005)	Asphalt (0.025)	Traffic Signals (0.003)	-	30 (0.008)	Use sidepath (bike not allowed) (0.013)
Fast Traffic	Tertiary (0.02)	Asphalt (0015)	Midblock (0.019)	Forest (0.005)	70 (0.008)	no bike lanes (0.026)
Missing Crossing	Residential (0.02)	Asphalt (0.022)	Three leg (0.023)	-	50 (0.013)	bike with track (0.014)
Missing Path	Tertiary (0.01)	Paved (0.004)	Midblock (0.05)	Forest (0.01)	50 (0.02)	shared lane (0.007)
People Behavior	Residential (0.023)	gravel (0.018)	Four leg (0.009)	Residential (0.006)	-	bike with track (0.013)
Poor Visibility	Tertiary (0.015)	ground (0.041)	Three leg (0.019)	Residential (0.006)	30 (0.015)	bike lane on both sides (0.007)
Unsafe Crossing	Primary (0.008)	-	Four leg (0.006)	-	-	-
Unsafe Path	Tertiary (0.015)	paved (0.005)	Midblock (0.014)	-	-	Bikelane with track (0.017)
Other	-	-	-	-	50 (0.014)	bidirectional bike lanes (0.005)

The findings show that surfaces such as cobblestones and paving stones are strongly linked to reports of bad road conditions in Antwerp and Gent. Particularly for cyclists, cobblestone roads are often considered hazardous. In Mol and Zonhoven, children are likelier to report dangerous intersections when the road surface is asphalt. Additionally, children are more likely to report issues related to fast traffic, missing crossings, or unsafe paths on roads with asphalt, concrete, or paving stones.

The model indicates that areas with forested land use significantly impact reporting fast traffic. In contrast, reports of people's behavior and poor visibility are more likely in residential areas.

Table 5. Average SHAP values of road features which has the highest positive effect on perceived risk in Zonhoven

	Road Type	Road Surface	Intersections	Land Use	Maximum Speed	Bike lanes
Bad Road	Tertiary (0.026)	gravel (0.011)	Midblock (0.047)	Industrial (0.005)	70 (0.009)	bikelane both sides (0.033)
Dangerous Intersection	Residential (0.012)	Asphalt (0.016)	Four leg (0.031)	Forest (0.005)	-	cycleway with track (0.016)
Fast Traffic	Residential (0.009)	paved (0.011)	Midblock (0.01)	Forest (0.01)	70 (0.012)	Bidirectional bikelane at right (0.004)
Missing Crossing	Primary (0.003)	Concrete (0.008)	Four leg (0.013)	Park (0.005)	70 (0.016)	cycleway with track left (0.005)
Missing Path	Residential (0.024)	Asphalt (0.011)	Midblock (0.019)	-	50 (0.009)	Bike designated (0.015)
People Behavior	Tertiary (0.01)	Concrete (0.018)	Midblock (0.019)	Residential (0.004)	30 (0.03)	Bike Lane on the right side (0.011)
Poor Visibility	Residential (0.019)	gravel (0.033)	Three leg (0.01)	Residential (0.005)	90 (0.018)	cycleway with track (0.023)
Unsafe Crossing	Tertiary (0.017)	Asphalt (0.008)	Three leg (0.014)	-	-	bike lane on both sides (0.013)
Unsafe Path	-	Unpaved (0.008)	-	-	-	Use sidepath (bike not allowed) (0.004)
Other	Residential (0.014)	Concrete (0.013)	-	Residential (0.002)	50 (0.005)	cyclestreet (0.014)

The model shows that roads with a speed limit of 30 km/h are strongly associated with reports of bad roads and people's behavior. Additionally, missing paths are more frequently reported on roads with a speed limit of 50 km/h. In Antwerp and Ghent, fast traffic is more likely to be reported on roads with a 30 km/h speed limit, while dangerous intersections are more commonly reported on roads with a 50 km/h speed limit. In Mol and Zonhoven, fast traffic is notably reported on roads with a speed limit of 70 km/h.

The model shows that roads designated as cycle streets or lacking bike lanes are more likely to be perceived as having bad roads in Antwerp and Ghent. Conversely, roads with bike lanes, particularly bidirectional bike lanes, are more likely to be perceived as bad roads in Mol and Zonhoven. Roads with bidirectional bike lanes, one-way bike lanes, or bike lanes on either side (or with dedicated bike tracks) are frequently associated with reports of missing crossings, people's behavior, poor visibility, unsafe crossings, or unsafe paths. Furthermore, roads with bike lanes on either side or bidirectional bike lanes in Antwerp, Gent, or Zonhoven are more likely to be associated with reports of fast traffic.

4 Discussion

The model in the current study suggests that traffic signals are often perceived as dangerous, particularly at intersections. According to a Federal Highway Administration (FHWA) study in the USA, approximately one-third of traffic fatalities occur at signal-controlled intersections [16]. Other research also found a link between traffic signals and children's injuries [17]. Several factors contribute to the safety risks associated with traffic signals, including their placement at hazardous intersections and, as reported by children and parents in the R2S descriptions, issues such as inadequate signal timing and non-compliance by other road users. Additionally, the current study reveals that children are more likely to report concerns about fast traffic, missing paths, and unsafe routes at midblock road segments not directly adjacent to intersections.

Several studies show that main and collector roads, compared to local roads, are related to children's traffic injuries [18, 19]. In the current study, main roads (primary and secondary) and collector roads (tertiary) positively impact perceiving unsafe crossings and paths. In addition, dangerous intersections are likely to be found on main roads in Antwerp and Mol. In contrast, based on the model, it will likely report fast traffic, missing crossings, and poor visibility on residential roads. This may be because most schools are on residential roads, which may relate to one study that shows schools located on local (residential) roads show a high collision rate compared to main roads [20]. Another study shows that parents perceive roads around schools as unsafe compared to main roads [21].

While residential roads are often considered safer due to reduced speed limits and lower traffic volumes [22], current study findings suggest that children and parents frequently report concerns about people's behavior and limited visibility in these areas. Conversely, participants reported issues with fast-moving traffic in forested regions, relatively commonly, potentially reflecting the generally higher speed limits in such zones.

A previous study found that children and parents perceive separate bike lanes as safer [23]. However, the current research indicates that cycle streets and roads without dedicated bike lanes are associated with fewer perceived bottlenecks. In contrast, most bottlenecks are more frequently reported on roads featuring bike lanes, bidirectional bike lanes, and bike lanes with dedicated tracks.

5 Conclusions

This study offers valuable insights into how children perceive road safety in diverse urban settings by integrating crowdsourced data from the Route2School (R2S) app with machine learning analysis. The findings highlight that traffic signals, particularly at intersections, are frequently perceived as unsafe. This perception is consistent with existing research, which has shown a high incidence of traffic fatalities and injuries at signal-controlled intersections. Factors contributing to this include suboptimal signal placement, inadequate signal timing, and non-compliance by other road users, all commonly reported by participants.

In addition to intersections, the analysis reveals that midblock road segment areas between intersections are also of concern, with frequent reports of fast-moving traffic, missing crossings, and unsafe paths. These findings emphasize the need for road safety strategies that extend beyond intersection improvements to include interventions in midblock areas, where vulnerabilities are often under-addressed.

Road hierarchy also plays a significant role in perceived safety. Main and collector roads (primary, secondary, and tertiary) were strongly associated with negative safety perceptions, particularly regarding unsafe crossings and paths. Interestingly, residential roads, while typically considered safer due to lower speed limits, were also linked to safety concerns, especially in areas surrounding schools. Participants reported issues such as poor visibility and unsafe behavior by other road users. These observations align with studies suggesting that local roads near schools may still pose considerable risks despite lower traffic volumes.

The study identifies a more nuanced picture of cycling infrastructure than previous findings. While earlier research has generally shown that separated bike lanes are perceived as safer, the current analysis found that reported bottlenecks were more frequent on roads featuring bike lanes, particularly bidirectional lanes, and protected tracks. However, this should not be interpreted as evidence that bike lanes are inherently unsafe. In Flanders, where cycling infrastructure is extensive and widely used, especially by children, the higher frequency of reported issues may partly reflect greater exposure rather than a lower level of safety. Moreover, the findings may point to localized design or maintenance issues or conflicts arising in shared or constrained spaces rather than problems with bike lanes. In contrast, cycle streets and roads without dedicated bike lanes were associated with fewer reported issues, possibly due to lower traffic volumes or more explicit rules of interaction. These findings underscore the need to evaluate cycling infrastructure not only by type but also by local context, design quality, and user experience.

The findings underscore the importance of incorporating subjective safety perceptions into road safety assessments. This study demonstrates a replicable approach for identifying safety concerns in urban environments by combining spatial data with crowdsourced insights and explainable machine learning models. Such an approach provides actionable information for urban planners and policymakers, enabling more targeted and context-sensitive interventions to improve the safety and mobility of vulnerable road users, particularly children.

Future research should explore additional contextual variables such as traffic volumes, environmental features, and temporal patterns. Longitudinal studies also help assess the impact of infrastructure changes over time, while comparative studies across different demographic groups extend the understanding of how various populations experience road safety.

References

1. Kyttä, M., Broberg, A., Haybatollahi, M., Schmidt-Thomé, K.: Urban happiness: context-sensitive study of the social sustainability of urban settings. Environ. Plann. B. Plann. Des. **43**(1), 34–57 (2015). https://doi.org/10.1177/0265813515600121
2. Carver, A., Timperio, A., Crawford, D.: Playing it safe: the Influence of Neighbourhood Safety on children's physical activity—a review. Health Place **14**(2), 217–227 (2008). https://doi.org/10.1016/j.healthplace.2007.06.004
3. Vias Institute, Road Safety Perceptions of Children and Parents, Brussels, Belgium, [Year]. https://www.vias.be/publications/Perceptie%20van%20de%20verkeersveiligheid/Road_Safety_Perceptions_of_children_and_parents.pdf
4. Nevelsteen, K., Steenberghen, T., Van Rompaey, A., Uyttersprot, L.: Controlling factors of the parental safety perception on children's travel mode choice. Accid. Anal. Prev. **45**, 39–49 (2012)
5. Amiour, Y., Waygood, E.O., van den Berg, P.E.: Objective and perceived traffic safety for children: a systematic literature review of traffic and built environment characteristics related to Safe Travel. Int. J. Environ. Res. Public Health **19**(5), 2641 (2022). https://doi.org/10.3390/ijerph19052641
6. Misra, A., Gooze, A., Watkins, K., Asad, M., Le Dantec, C.A.: Crowdsourcing and its application to transportation data collection and management. Transp. Res. Rec. J. Transp. Res. Board **2414**(1), 1–8 (2014). https://doi.org/10.3141/2414-01
7. "Route2School," Route2school.be (2024). https://www.route2school.be/. Accessed 25 Aug 2024
8. OpenStreetMap Contributors, "OpenStreetMap," OpenStreetMap (2023). https://www.openstreetmap.org
9. Pocock, T., Moore, A., Keall, M., Mandic, S.: Physical and spatial assessment of school neighbourhood built environments for active transport to school in adolescents from Dunedin (New Zealand). Health Place **55**, 1–8 (2019). https://doi.org/10.1016/j.healthplace.2018.10.003
10. Lee, G., Park, Y., Kim, J., Cho, G.-H.: Association between intersection characteristics and perceived crash risk among school-aged children. Accid. Anal. Prev. **97**, 111–121 (2016)
11. Alonso, F., Esteban, C., Tortosa, F., Useche, S.: Perception of road safety in children's environment. Am. J. Educ. Res. **5**(3), 273–278 (2017). https://doi.org/10.12691/education-5-3-7
12. "Road accidents | Statbel," Fgov.be, 25 June 2024. https://statbel.fgov.be/en/themes/mobility/traffic/road-accidents
13. "Open Data | Statbel," Fgov.be (2021). https://statbel.fgov.be/en/open-data?category=23. Accessed 04 March 2025
14. Breiman, L.: Random forests. Mach. Learn. **45**(1), 5–32 (2001). https://doi.org/10.1023/A:1010933404324

15. Lundberg, S., Lee, S.-I.: A unified approach to interpreting model predictions. In: Proceedings of the 31st International Conference on Neural Information Processing Systems (NeurIPS), Long Beach, CA, USA, pp. 4765–4774 (2017)
16. Federal Highway Administration, "About Intersection Safety | FHWA," Dot.gov (2015). https://highways.dot.gov/safety/intersection-safety/about
17. Rothman, L., Macarthur, C., To, T., Buliung, R., Howard, A.: Motor vehicle-pedestrian collisions and walking to school: the role of the built environment. Pediatrics **133**(5), 776–784 (2014). https://doi.org/10.1542/peds.2013-2317
18. Cloutier, M.-S., Apparicio, P., Thouez, J.-P.: GIS-based spatial analysis of child pedestrian accidents near primary schools in Montréal, Canada. Appl. GIS **3**(4), April 2007. https://doi.org/10.4225/03/57E9B71457976
19. Yu, C.-Y.: How differences in roadways affect school travel safety. J. Am. Plann. Assoc. **81**(3), 203–220 (2015). https://doi.org/10.1080/01944363.2015.1080599
20. McArthur, A., Savolainen, P.T., Gates, T.J.: Spatial analysis of child pedestrian and bicycle crashes. Transp. Res. Rec. J. Transp. Res. Board **2465**(1), 57–63 (2014). https://doi.org/10.3141/2465-08
21. Rothman, L., et al.: Associations between parents' perception of traffic danger, the built environment and walking to school. J. Transp. Health **2**(3), 327–335 (2015). https://doi.org/10.1016/j.jth.2015.05.004
22. Rothman, L., et al.: School environments and social risk factors for child pedestrian-motor vehicle collisions: a case-control study. Accid. Anal. Prev. **98**, 252–258 (2017). https://doi.org/10.1016/j.aap.2016.10.017
23. Hopkins, D., Mandic, S.: Perceptions of cycling among high school students and their parents. Int. J. Sustain. Transp. **11**(5), 342–356 (2016). https://doi.org/10.1080/15568318.2016.1253803

Setting Priorities for Safety Improvements on Two-Lane Roads in Israel

Victoria Gitelman(✉)

Transportation Research Institute, Technion, Haifa, Israel
trivica@technion.ac.il

Abstract. This study aimed to establish priorities for advancing infrastructure improvements on the two-lane interurban road network to enhance road safety. For that purpose, the safety level of all road sections was evaluated using a Bayesian approach and a cost-effectiveness analysis applied to assess potential safety benefits from the implementation of four infrastructure measures, such as: upgrading to a dual-carriageway, setting a central barrier, adding rumble strips on centrelines and rumble strips on edge-lines of the roadway. The lists of high-priority sections that were produced by the study are associated with the highest potential accident savings, on the road network, stronger cost-effectiveness ratios and a geographic match of sections to enable consistent infrastructure changes along the roads. The priority lists outline a program of future activities on improving the safety level of two-lane roads on the national road network and thus may support a systematic work of the relevant authorities.

Keyword: Road safety · Road infrastructure · Two-lane roads · Infrastructure measures · Accident savings · Cost-effectiveness

1 Introduction

Recognizing the global burden of road traffic injury around the world [1] and the critical role of road infrastructure in road accident occurrences and their outcomes, substantial efforts are invested today in providing safer road infrastructure [2–4]. Regarding the non-urban road network, in many countries, two-lane (or single-carriageway) roads are in focus as they are characterized by the lowest safety level compared to other road types. This situation arises from the increased risk of head-on collisions and generally lower design standards of two-lane roads compared to others [2, 5, 6].

For example, in Israel, roads of higher hierarchy - motorways and dual-carriageways, are built to carry large traffic volumes and thus feature higher design standards, e.g. more travel lanes, wide lanes and shoulders, larger curve radii, forgiving roadside conditions, physical separation between traffic directions [7]. In contrast, two-lane roads are intended for lower traffic volumes and thus have narrower lanes and shoulders, lacking physical separation between opposing traffic flows, etc. As a result, two-lane roads are associated with a much higher accident risk per vehicle-km traveled.

The Netivey Israel Company (NIC) is responsible for the construction and maintenance of the majority of the non-urban road network of the country, which currently

E. Macioszek et al. (Eds.): TSTP 2025, LNNS 1769, pp. 16–29, 2026.
https://doi.org/10.1007/978-3-032-14094-4_2

accounts for over 6,500 km. Recent estimates showed that the annual average of accident numbers on the non-urban network was about 3,900 injury accidents, including 170 fatal and around 660 serious accidents (4% and 17%). The NIC network length subdivision includes 75% of two-lane roads, 20% of dual-carriageways, and 5% of motorways, where severe accidents rates per 100 million vehicle-km traveled were 3.0, 1.4 and 1.0, respectively. Furthermore, the models developed for predicting accidents on Israeli roads showed that for any level of exposure, the safety performance of two-lane roads is worse compared to other road types [8].

Being aware of the higher accident risk of two-lane roads, which comprise the majority of the non-urban network, the National Road Safety Authority (NRSA) initiated a study aiming to establish priorities for advancing infrastructure improvements on two-lane roads that will help enhance road safety. It was stated that the study methods should follow the scientific rules of road safety evaluations [5, 9, 10], while the resulting high-priority lists should be flexible to support practical work.

2 Methodology

The methods and tools applied in this study assessed the safety level of two-lane road sections and the potential safety benefits resulting from their infrastructure improvements. The latter stems from the expected reduction in accident numbers due to the implemented infrastructure measures.

2.1 The Study Data

The evaluations referred to the two-lane road network managed by the NIC, and used accident data from three recent years, 2021–2023. The initial information on road infrastructure, traffic exposure and accident data, for all road segments, was received from the NIC database. The NIC data are originally collected through road field surveys, for segments of about 100 m in length. For the study analyses, more essential section lengths were needed, to reflect the network subdivision between major road intersections. Thus, a generalization was conducted, where segments from the same road and with similar traffic volumes were combined in one road section where the section ends were defined by main intersections. This way, the original 3,325 segments, of 4,199 km in length in total, were converted into 1,221 road sections.

2.2 Infrastructure Measures Examined

To improve the safety conditions of two-lane roads, various infrastructure solutions can be found in the literature [5, 6, 11–13], including: improvements to the road cross-section, both horizontal and vertical; shoulder enhancements; providing forgiving roadsides, by setting vehicle recovery zones or installing safety barriers; setting a soft median between travel directions, in the form of rumble strips or a flexible divider; adding a central barrier with creating a "1 + 1" cross-section or a "2 + 1" cross-section; and finally, upgrading to a dual-carriageway.

Since infrastructure data in the NIC database were limited and did not include reliable values for road layout characteristics of each segment, it was not possible to conduct evaluations for some infrastructure measures, such as widening lanes or shoulders, improving curvature, or creating forgiving roadsides. In addition, a "2 + 1" road layout is not common in Israeli practice, while "1 + 1" cross-section solutions were applied in the past decades [13]. Hence, the study examined the potential safety impacts of four types of infrastructure measures on two-lane roads, which are: upgrading to a dual-carriageway; setting a central barrier ("1 + 1" layout); adding rumble strips on the centreline, and adding rumble strips on edge-lines of the roadway.

2.3 Methods and Tools for Conducting Safety Evaluations

In line with the requirements for correct road safety evaluations [5, 6, 10, 14, 15], the estimate of the safety level of a site should aspire to assess the typical number of accidents for that site, which reflects a multi-year accident expectancy. Such an estimation should account for the phenomena of random nature of accidents, traffic exposure, regression-to-the mean and accident severity, and should preferably apply an empirical Bayes evaluation. By this method, the typical (or *expected*) number of accidents at a site is estimated as a weighted value of two components: (a) the mean number of accidents that are predicted for such type of sites (e.g., road sections with similar layouts and traffic volumes); and (b) the number of accidents actually observed at this site. The first component should preferably be estimated using a statistical model known as safety performance functions (SPFs), fitted to local conditions [6, 10, 15].

An SPF is a multivariate model, which shows a relationship between accidents and traffic flows and (optionally) other road characteristics of the road sites. As the form of the relationship may differ for various accident severities (slight, serious, fatal), whereas the latter are also associated with different economic costs, it is advisable to develop SPFs for various accident severity levels. Finally, "black-spots" on the road network can be recognized using a high positive difference between the number of accidents *expected* at the site and that *predicted* for similar sites using the SPF, which are weighted across various severity levels [9, 10].

The Safety Management System (SMS) of the NIC adopted a Bayesian approach for evaluation of safety performance of various road sites, and SPFs for local road types were developed by previous research [8]. In the current study, to assess the safety level of two-lane road sections and the safety benefits from upgrading to a dual-carriageway, the following tools, from the SMS, were applied:

1. Estimating the typical number of accidents for a road section

The expected number of accidents at a site is estimated by combining the number of accidents observed at the site and those predicted using the SPFs, as follows:

$$m_i = (w_i \cdot SP_i) + (1 - w_i) \cdot \frac{X_i}{n},\ w_i = \frac{k_i}{\left(k_i + \left(n \cdot \frac{SP_i}{L}\right)\right)} \quad (1)$$

where: m_i - the expected number of accidents at the site; n – the number of years for counting accidents at the site; X_i – the number of accidents observed at the site in n

years; SP_i – the predicted number of accidents for this type of sites (using SPFs); k_i – the estimate of uncertainty of the SPF model; L – length of the road section, in km. The values are obtained for the three levels of accident severity, where $i = L,S,F$ indicate slight, serious and fatal accidents, respectively.

The *weighted* typical number of accidents at the site is estimated as a weighted sum of the values across all severity levels:

$$w\_m = \sum_i (m \cdot \lambda_i \tag{2}$$

where: λ_i – the weights based on the economic costs of typical accidents. Based on the guidelines on the appraisal of transport projects [16], the estimations for local conditions demonstrated [8] that the average costs of fatal, serious and slight injury accidents on non-urban road sections were 8.16, 1.236 and 0.134 million NIS (New Israeli Shekels, 1 US$ ~ 3 NIS), thus providing a ratio of 61:9:1, which defined the weights above.

2. Statistical models for predicting the number of accidents on a road section (SPFs)

Among the models developed for the SMS [8], in the current study evaluations, models for two road types were used, i.e. for two-lane and dual-carriageway roads. A two-lane road has one travel lane in each direction, without a physical separation between them. A dual-carriageway road has a physical separation between the carriageways, in the form of a built median and/or safety barrier installed, and 2–3 lanes in each travel direction. The notations used in the models are: SP_L, SP_S, SP_F - the number of slight, serious and fatal accidents expected on the section, per year; L - the length of road section, in km; V – average daily traffic volume, in thousands of vehicles; e – exponent. The SPFs for a two-lane road section are as follows:

$$SP_L = L \cdot e^{-3.533} \cdot V^{0.645} \cdot \max\left\{1, \left(\frac{V}{6}\right)\right\}^{0.724}$$

$$SP_S = L \cdot e^{-5.205} \cdot V^{0.947}$$

$$SP_F = L \cdot e^{-6.109} \cdot V^{1.041}$$

where the estimates of the models' uncertainty (for parameter k_i in Eq. (1)) are: 0.475, 0.450 and 0.90 for slight, serious and fatal accidents, respectively.

The SPFs for predicting accidents on a dual-carriageway road section have the form:

$$SP_L = L \cdot e^{-4.314} \cdot V^{1.111} \cdot \max\left\{1, \left(\frac{V}{40}\right)\right\}^{-1.172}$$

$$SP_S = L \cdot e^{-7.931} \cdot V^{1.572} \cdot \max\left\{1, \left(\frac{V}{40}\right)\right\}^{-1.918}$$

$$SP_F = L \cdot e^{-4.206} \cdot V^{0.202} \cdot \max\left\{1, \left(\frac{V}{40}\right)\right\}^{1.019}$$

where the estimates of the models' uncertainty are 0.588, 0.477 and 0.474, for slight, serious and fatal accidents, respectively.

It should be noted that for both road types, the SPFs show an increase in accidents of all severity levels when the traffic volume increases, yet the relationships are not monotonous: on two-lane roads, for slight accidents, the accident growth rate strengthens as traffic volumes exceed 6,000 vehicles per day, while on dual-carriageways, for traffic volumes over 40,000 vehicles per day, the accident rate is moderated for slight and serious accidents, while for fatal accidents it becomes stronger [8].

3. Calibration factors

In addition, as the SPFs for local conditions were developed based on accident data from the previous decade, while accident numbers on the network change overtime, for the current study the SPF results were calibrated based on examining accident data recorded on the road network in recent years. Following [17], for each model, calibration factors were estimated when the sum of actual accident numbers observed across all segments in the network was related to the sum of predicted accident numbers on those segments. The calibration factors obtained were: on two-lane roads – 1.049, 1.672, 0.890, on dual-carriageways – 1.070, 1.868, 0.696, for slight, serious, and fatal accidents, respectively. They were used to adjust the predicted values from the models SP_L, SP_S, SP_F, respectively.

2.4 Estimating Accident Savings and Setting Priorities

The study focused on safety impacts of the infrastructure measures, without mobility considerations. Hence, the appropriate assessment framework was a cost-effectiveness (CE) analysis, where a CE ratio estimates the number of accidents prevented by the measure related to implementation costs [18, 19]. In this study, the CE ratio nominator was the expected reduction in accidents following the implementation of the measure, which, in most cases, was estimated by multiplying the weighted number of accidents at the site (see Eq. (2)) with the accident reduction factor of the measure. Accident reduction factors of the measures, also known as safety effects of road infrastructure improvements, are based on previous road safety research; for the current study, the values of accident reduction factors were selected from major international sources of knowledge on the topic, e.g. [5, 6, 11, 20, 21] and local evaluation studies, e.g. [13].

In the case of upgrading to a dual-carriageway, the estimation of the number of accidents prevented was different. For each road section, with a given traffic volume and length, the predicted numbers of accidents were evaluated for a dual-carriageway layout. Then, differences between the expected numbers of accidents on the section being two-lane and the predicted numbers of accidents for the same section with a dual-carriageway layout, were estimated and weighted across for three severity levels, using the weights from Eq. (2). The result provided an estimate of the weighted number of accidents to be saved due to the measure. Evidently, only sections with positive values of this estimate are to be considered for safety benefits from upgrading to a dual-carriageway.

The denominator of the CE ratio was obtained by multiplying the unit costs of the measure by section length. The unit costs of implementation were assigned based on the NIC price-list for road construction and maintenance, and the cost-estimates reported by the NIC for treatments of hazardous locations on the road network. Table 1 provides a summary of values applied in this study for CE evaluations of infrastructure measures on the two-lane road sections.

To set treatment priorities, road sections can be ranked using the estimation results, in accordance with the level of section un-safety (number of accidents expected), number of accidents saved by the measure or CE ratios, whereas higher values of each parameter indicate the sites with higher safety benefits from the treatment. Several types of ranking were attempted in the study and, finally, a combined approach was applied to enable flexibility of findings.

Table 1. Accident reduction factors and unit costs of road infrastructure improvements on two-lane road sections.

Infrastructure measure	Accident reduction factor		Unit costs of implementation, million NIS
	Value applied in the study	Sources of estimates	
Setting a central barrier ("1 + 1")	–40%	[6, 11–13, 20, 21]	6.0
Rumble strips on centreline	–15%	[5, 6, 11, 20]	0.8
Rumble strips on edge-lines	–20%	[5, 11, 20]	0.04
Upgrading to a dual-carriageway	Not applicable	*	8.0

Notes: NIS - New Israeli Shekels, currently 1 US$ ~ 3.5 NIS. * The number of accidents saved is estimated based on differences between the expected accident numbers on a two-lane road and predicted accident numbers on a dual-carriageway road, for the same section.

3 Results

3.1 Evaluation Examples

Table 2 provides examples of evaluating two-lane road sections, including section characteristics (location, length, traffic volume), numbers of accidents observed in the past years, the numbers expected for a two-lane road and predicted for a dual-carriageway layout, as well as the numbers of accidents to be saved following each measure's implementation, with a CE ratio value.

It can be noted that on all road sections shown in Table 2 no low accident numbers were observed, non-negligible accident numbers are expected annually with the two-lane layout and tangible safety benefits can be attained from each measure implemented. Among the cases presented, the road 784 section is longer than others and with higher traffic volumes, and thus is associated with higher annual accident frequencies and more accidents saved following the treatments. Yet, the CE ratio values are close among the sections, for all treatment types.

Table 2. Examples of evaluating two-lane road sections, in the study.

Road	Section characteristics: km start	km end	length, km	AADT	Accidents observed*: total	fatal	serious	slight
4	243.6	246.1	2.5	17,799	4.00	0.00	0.67	3.33
784	11.2	18.6	7.4	19,028	8.33	0.67	1.00	6.67
89	52.4	54.7	2.2	12,786	3.00	0.33	0.67	2.00

Road	Accidents expected#: m_L	m_S	m_F	$w\text{-}m^{**}$	Accidents predicted##: SP_L	SP_S	SP_F
4	2.73	0.51	0.09	12.8	0.58	0.16	0.07
784	5.87	1.05	0.35	36.7	1.82	0.51	0.21
89	1.49	0.41	0.09	10.6	0.35	0.08	0.06

	Accidents** saved per year (1) and a CE ratio (2) for treatment:							
	rumble strips on centreline		rumble strips on edge-lines		setting a central barrier ("1+1")		upgrading to a dual-carriageway	
Road	(1)	(2)	(1)	(2)	(1)	(2)	(1)	(2)
4	1.9	0.95	2.6	25.2	5.1	0.34	6.4	0.31
784	5.5	0.93	7.3	24.8	14.7	0.33	17.2	0.29
89	1.6	0.89	2.1	23.8	4.2	0.32	5.9	0.33

Notes: AADT – average annual daily traffic; *a 3-year average; #on a two-lane road, per year, see eq. (1); **weighted accident numbers, per year, see eq. (2); ##for a dual-carriageway road, per year.

3.2 Sections Preferred for Measures' Implementation

As indicated in Sect. 2, selecting road sections which are preferable for treatment, i.e. with the highest safety benefits for the road network, can be based on various estimates, e.g., the weighted number of accidents expected in the current state, the number of accidents annually saved by the treatment or a CE ratio per treatment, where higher values of any indicator promise higher safety benefits. Initially, several rankings were tried on the road network but did not provide satisfactory results. For example, a ranking based on the typical number of accidents expected (as normalized per 1 km of section length) was prepared and indeed indicated sections with the highest density of weighted accidents in the current state, but this ranking suffered from highlighting short sections and was not helpful in selecting preferable measures. Another attempt of ranking used

CE ratio values while considering all types of measures for all road sections together; however, the result demonstrated that the top of the list was occupied by the measure of "rumble strips on edge-lines" for the majority of road network as this treatment has substantially lower costs relative to others. Furthermore, in working discussions with the NRSA steering team, besides the priorities based on safety benefits, the need to consider neighboring sections was raised because in practice road infrastructure layouts should be consistent along the road and hence road infrastructure improvements should be applied on consecutive road sections.

Finally, to define treatment priorities on the road network, a combined approach was suggested, which considers each measure separately, ranks sections in accordance with their safety benefit potential and in selecting preferred sections also accounts for section neighbors on the same road. In addition, it supports the selection of high-priority sections by presenting absolute and relative safety benefits, both for separate sites and the whole list.

More specifically, regarding each measure (out of the four examined in the study, see Table 1), a ranking of the road sections was performed according to the extent of accident reduction resulting from the measure implementation (the weighted number of accidents saved per year). This allowed pointing to the sections with the highest potential for accident savings associated with the given infrastructure change. Then, to identify preferred sections for the treatment, two considerations were combined – high safety benefit from improving the site and creating a continuous sequence of sections for the treatment as is required for actual implementation on the road. To satisfy the second consideration, 100 top-ranked sites were filtered and classified into the "top 20", which composed the base for selection (*level A*), while the next 40 were marked as *level B*, and the last 40 as *level C*. For each of *level A* sections, sites from levels B and C were attached to compose a group if a continuous kilometer sequence was observed along the road. Furthermore, the whole road network was checked to ensure that all sections adjacent to each *level A* section were examined. (Due to substantially lower costs, an exception was made for measure "rumble strips on edge-lines", where the selection was based on the 200 top-ranked sites, which were subdivided into "top-100", next 50 and last 50).

For the received list of the preferred sections, cumulative estimates were provided, including: accident savings per year, treatment costs, total length of sections for the treatment, and a CE ratio. In addition, for each separate section an estimate of the number of accidents to be saved and a CE value of the treatment were kept. This way, both absolute and relative safety benefits are presented in the final list of the preferred sections. It should be emphasized that the final list can be "stopped" on any line (road section) according to the budget available for treatment. It is also possible to "skip" a certain section or group of sections on the road, for example, due to too great length (and hence, substantial budget to be spent on the treatment). Thus, the final list enables flexibility in conducting the practical work.

Table 3 shows an example of the final list of high-priority two-lane sections for the measure "setting a central barrier". The list includes 34 sections (in 17 groups of roads) with a total length up to 730 km, total savings up to 793 accidents and total costs up to 4,382 million NIS. Figure 1 illustrates the cumulative estimates of this measure, indicating an increase in the total accidents saved along with a stronger rise in the total costs when the total length of treatment increases. The CE ratio values are highest, over 0.40, for the first 48 km of treatment, remain over 0.30 for the initial 300 km, and decrease to 0.18 at the end of the list.

Table 4 provides a summary of cumulative estimates of the final lists for all the measures examined. Figure 2 illustrates the cumulative estimates for the other three measures, in addition to Fig. 1.

Table 3. Final list of high-priority sections for measure "setting a central barrier".

			Section estimates		Cumulative estimates			
Road	Km start	Km end	Accidents to be saved*	CE ratio	Accidents to be saved*	Costs, million NIS	CE ratio	Treatment length, km
781	10.2	16.8	23.7	0.60	24	40	0.60	6.6
562	0.0	7.5	19.8	0.44	44	85	0.51	14.2
65	54.5	58.1	8.9	0.41	53	107	0.49	17.8
65	61.3	69.4	20.3	0.42	73	155	0.47	25.9
806	0.0	8.5	19.9	0.39	93	206	0.45	34.4
675	0.0	9.0	20.5	0.38	113	260	0.44	43.3
60	13.5	17.8	9.5	0.37	123	285	0.43	47.6
60	18.2	49.8	54.6	0.29	177	475	0.37	79.2
60	49.8	60.3	30.0	0.47	207	538	0.39	89.7
60	65.5	68.7	9.2	0.48	217	557	0.39	92.9
55	13.0	32.2	37.0	0.32	254	673	0.38	112.1
60	80.7	82.8	11.6	0.92	265	685	0.39	114.2
60	95.9	97.7	5.9	0.57	271	696	0.39	115.9
60	97.7	99.4	2.1	0.20	273	706	0.39	117.7
60	99.4	105.0	7.9	0.24	281	740	0.38	123.3
60	105.0	135.5	51.6	0.28	333	923	0.36	153.8
60	135.5	153.2	35.4	0.33	368	1,029	0.36	171.5
60	153.2	160.3	8.2	0.19	376	1,071	0.35	178.6
446	8.4	13.1	8.6	0.31	385	1,099	0.35	183.2
446	13.1	30.6	29.3	0.28	414	1,204	0.34	200.7
505	37.1	75.7	61.1	0.26	475	1,436	0.33	239.3
854	0.0	14.2	20.4	0.24	496	1,521	0.33	253.5

(*continued*)

Table 3. *(continued)*

			Section estimates		Cumulative estimates			
Road	Km start	Km end	Accidents to be saved*	CE ratio	Accidents to be saved*	Costs, million NIS	CE ratio	Treatment length, km
293	0.0	17.4	23.8	0.23	519	1,626	0.32	271.0
25	106.6	135.3	28.7	0.17	548	1,798	0.30	299.7
80	176.4	196.4	20.1	0.17	568	1,918	0.30	319.7
232	15.5	47.4	25.9	0.14	594	2,110	0.28	351.6
90	69.3	207.1	59.4	0.07	653	2,936	0.22	489.4
90	207.1	210.1	2.0	0.11	655	2,954	0.22	492.4
90	210.1	258.7	16.5	0.06	672	3,246	0.21	541.0
90	258.8	363.8	71.9	0.11	744	3,877	0.19	646.1
90	363.8	365.4	1.6	0.18	745	3,886	0.19	647.6
90	366.0	374.4	8.7	0.17	754	3,937	0.19	656.1
40	73.4	120.2	20.3	0.07	774	4,217	0.18	702.9
40	120.2	147.5	19.1	0.12	793	4,382	0.18	730.3

Notes: * weighted accident number, per year; NIS - New Israeli Shekels.

It can be seen that, overall, the high priority lists include 30–40 road sections (in 17–18 road groups) for more expensive treatments and a longer list of 138 sections (in 78 groups) for "rumble strips on edge-lines". All the lists were continued for substantial total costs, to outline a long-term perspective for intervention, more than 1–2 years. As the measures differ in their unit costs, the total length for treatment varies widely, from 347 km for upgrading to a dual-carriageway to over 1,700 km for rumble strips on edge-lines. Similarly, the CE ratio values differ substantially between the treatments evaluated. Yet, across all the measures, the most effective interventions will be on the first 50–140 km of the priority lists.

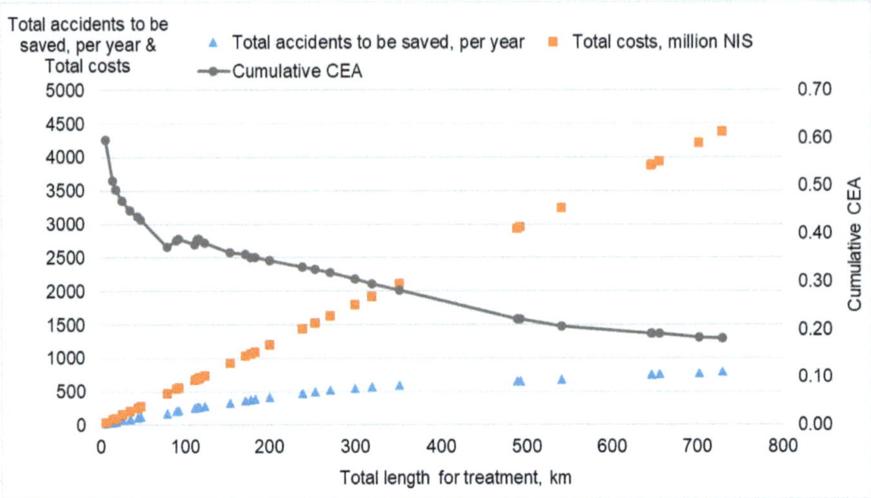

Fig. 1. Cumulative estimates of the list of high-priority sections for measure "setting a central barrier"

Table 4. Summary of cumulative estimates of the lists of high-priority road sections for all measures examined.

Measure	No of sections in the list (No of road groups)	Total length, km	Accidents to be saved, per year	Total costs, million NIS	CE ratio range	Treatment length for highest CE values
Setting a central barrier	34 (17)	730	793	4,382	From 0.60 to 0.18	48 km, CE > 0.40
Rumble strips on centreline	29 (17)	715	286	572	From 1.68 to 0.50	140 km, CE > 1.0
Rumble strips on edge-lines	138 (78)	1,765	872	71	From 69 to 12	123 km, CE > 30
Upgrading to a dual-carriageway	40 (18)	347	728	2,776	From 0.65 to 0.26	79 km, CE > 0.40

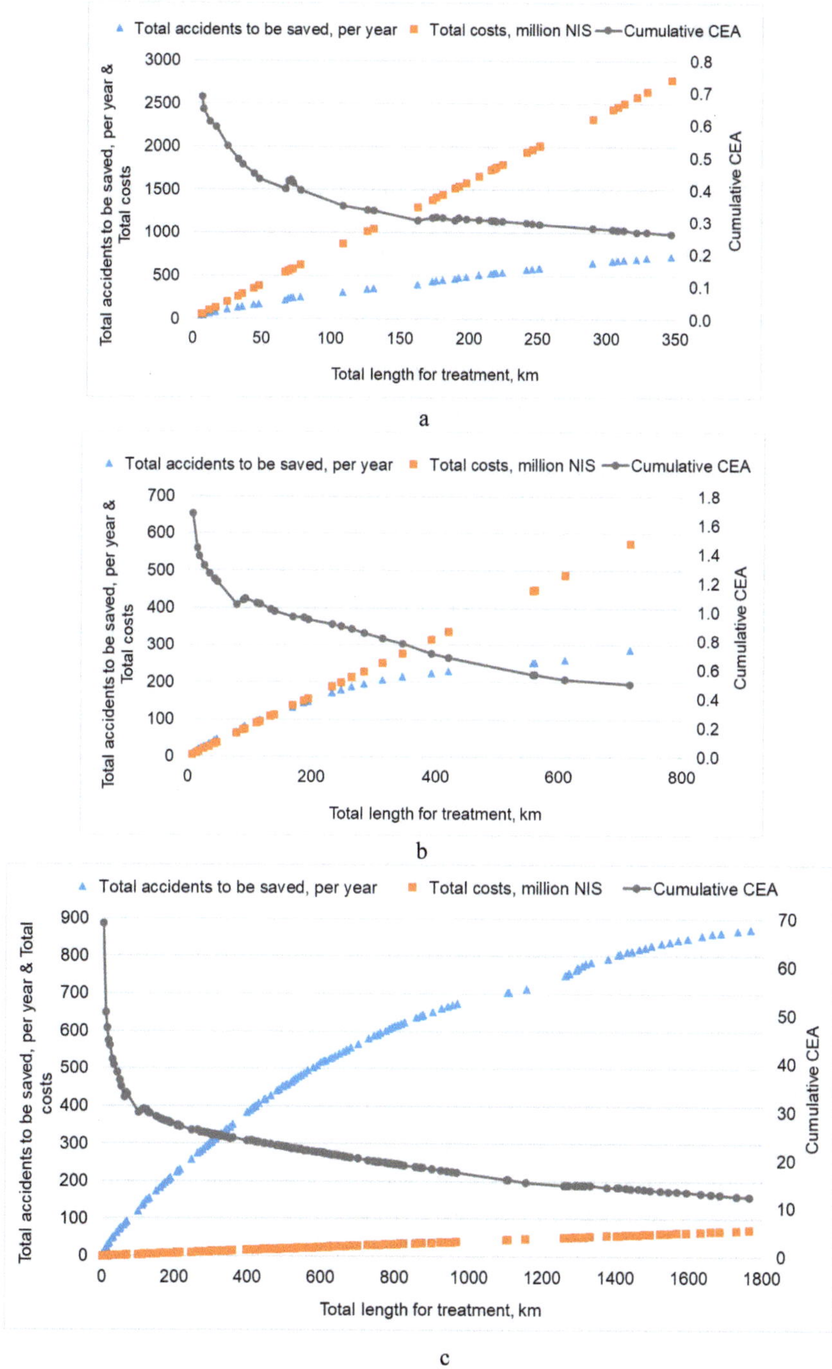

Fig. 2. Cumulative estimates of the lists of high-priority sections for other measures: a - upgrading to a dual-carriageway; b - adding rumble strips on centrelines; c - adding rumble strips on edge-lines

4 Conclusions

Recognizing the worse safety level of the two-lane roads on the non-urban road network when compared with dual-carriageways and motorways [2, 5, 6, 8], in this study a scientific approach was developed for selecting road sections for conducting infrastructure improvements. The measures considered for implementation were with proven safety effects as found by previous international and local evaluation research [5, 11–13, 20, 21] and included four infrastructure changes on two-lane road sections: upgrading to a dual-carriageway, setting a central barrier, adding rumble strips on centrelines and adding rumble strips on edge-lines of the roadway.

The method for selecting high-priority sections for each treatment, in this study, was based on the evaluations of safety performance of road sites which satisfied the requirements for correct road safety evaluations [5, 6, 10, 14, 15], and were applied in combination with a cost-effectiveness assessment of road safety measures [18, 19]. Such a combination of a Bayesian approach with a cost-effectiveness evaluation, on a national scale, is not common yet in preparation of road safety interventions for road network [22, 23].

In particular, the presented methodology for setting priorities for treatment of two-lane roads, on the national road network in Israel, was developed and applied for the first time. In addition, when conducting the study evaluations, a geographic match of the preferred section lists was conducted to stimulate consistent infrastructure changes along the roads and enable flexibility in actual implementation.

The priority lists supplied by the study outline a program of future activities on improving the safety level of two-lane roads on the national road network and thus may support a systematic work of the authorities, e.g. the NRSA and the NIC, in the field. Furthermore, the study findings may contribute to the development of a Vision zero program and other nation-wide safety interventions in the country which are promoted today by leading international bodies [2, 3, 24].

Acknowledgements. The study was commissioned and supervised by the Israeli National Road Safety Authority.

References

1. Global Status Report on Road Safety 2023. World Health Organization, Geneva (2023)
2. International Transport Forum: Zero Road Deaths and Serious Injuries: Leading a Paradigm Shift to a Safe System. OECD Publishing, Paris (2016)
3. European Union Road Safety Policy Framework 2021-30 – Next steps towards 'Vision Zero'. European Commission, Directorate-General for Mobility and Transport, Brussels (2020). https://doi.org/10.2832/391271
4. Papadimitriou, E., Filtness, A., Theofilatos, A., Ziakopoulos, A., Quigley, C., Yannis, G.: Review and ranking of crash risk factors related to the road infrastructure. Accid. Anal. Prev. **125**, 85–97 (2019)
5. Elvik, R., Hoye, A., Vaa, T., Sorensen, M.: The Handbook of Road Safety Measures, 2nd Edition. Emerald Group Publishing (2009)

6. Highway Safety Manual. American Association of State Highway and Transportation Officials, Washington, DC (2010)
7. Guidelines for Geometric Design of Roads. Ministry of Transport, Jerusalem, Israel (2018)
8. Gitelman, V., Doveh, E.: Development of supporting tools for safety management of non-urban roads, in Israel. In: 21st International Conference on Transport Science (ICTS 2024), Transportation Research Procedia, vol. 83, pp. 504–512 (2025)
9. Persaud, B., Lyon, G., Nguyen, T.: Empirical Bayes procedure for ranking sites for safety investigation by potential for safety improvement. Transp. Res. Rec. **1665**, 7–12 (1999)
10. World Roads Association (PIARC): Road Safety Manual. A manual for practitioners and decision makers on implementing safe system infrastructure (2025). https://roadsafety.piarc.org/en
11. European Road Safety Decision Support System. www.roadsafety-dss.eu
12. Austroads: Effectiveness of Road Safety Engineering Treatments. Austroads Publication AP-R422-12. Sydney, Australia (2012)
13. Gitelman, V., Carmel, R., Pesahov, F.: The evaluation of safety efficiency of non-urban infrastructure improvements; a case-study. Eur. Transp. Res. Rev. **6**, 477–491 (2014)
14. Hauer, E.: Observational Before-After Studies in Road Safety. Pergamon (1997)
15. Bahar, G., Masliah, M., Mollett, C., Persaud, B.: Integrated Safety Management Process. National Cooperative Highway Research Program Report 501, Transportation Research Board, Washington, DC (2003)
16. Guidelines on Feasibility Assessment of Transport Projects. Ministry of Transport and Ministry of Finance, Jerusalem, Israel (2012)
17. Elvik, R.: A before-after study of the effects on safety of environmental speed limits in the city of Oslo Norway. Safety Sci. **55**, 10–16 (2013)
18. Yannis, G., Gitelman, V., Papadimitriou, E., Hakkert, A.S., Winkelbauer, M.: Testing a framework for the assessment of road safety measures. Transp. Rev. **28**(3), 281–301 (2008)
19. A Guide to the Economic Efficiency Analysis (E3) Calculator. The SafetyCube Road Safety Decision Support System (2018). https://www.safetycube-project.eu
20. Harkey, D.L., Srinivasan, R., Baek, J., et al.: Accident modification factors for traffic engineering and ITS improvements. National Cooperative Highway Research Program Report 617, Transportation Research Board, Washington, DC (2008)
21. Austroads: Road Safety Engineering Risk Assessment. Part 6: Crash Reduction Factors. Austroads Technical Report AP-T151/10. Sydney, Australia (2010)
22. Sorensen, M., Elvik, R.: Black spot management and safety analysis of road network - Best practice guidelines and implementation steps. RiPCORD-iSEREST project Deliverable D6 (2008)
23. International Transport Forum: Road Infrastructure Safety Management. OECD Publishing, Paris (2015)
24. Global Plan for the Decade of Action for Road Safety 2021–2030. World Health Organization, Geneva (2021). www.who.int/publications/m/item/global-plan-for-the-decade-of-action-for-road-safety-2021-2030

Not Just the Driver: How Road Design Influences Speeding and Traffic Safety

Marko Šoštarić, Marijan Jakovljević, Juraj Leonard Vertlberg, and David Gruhonjić(✉)

Faculty of Transport and Traffic Sciences, University of Zagreb, Zagreb, Croatia
{msostaric,mjakovljevic,jvertlberg,dgruhonjic}@fpz.unizg.hr

Abstract. Road traffic accidents remain a leading cause of death and injury worldwide, with speed recognized as a critical risk factor. This paper examines the relationship between road design, operating speed, and accident occurrence, using both Croatian case studies and international best practices. Findings reveal that mismatches between road geometry, design speed, and posted speed limits often encourage drivers to travel above safe speeds, undermining enforcement efforts and increasing crash risk. Case studies such as the Turanj Bypass illustrate how inadequate infrastructure, insufficient signage, and lack of self-explaining and forgiving road principles contribute to unsafe driving behavior. The paper highlights the need for a systemic approach, recommending the adoption of modern design philosophies that intuitively guide driver behavior and mitigate the consequences of human error to achieve substantial and sustainable improvements in road safety.

Keywords: Road safety · speed management · road design · self-explaining roads · forgiving infrastructure

1 Introduction

Road accident fatalities and severe injuries remain among the most pressing global challenges, posing significant threats to public health and safety. According to the World Health Organization [1], approximately 1.35 million people lose their lives in road traffic accidents worldwide every year, with an even greater number sustaining injuries. Traffic accidents are currently the eighth leading cause of death globally and the leading cause of death for individuals aged 5 to 29 years [2]. The burden is particularly severe in low- and middle-income countries, where deficiencies in road infrastructure, vehicle fleets, and road safety education contribute to disproportionately high rates of fatalities and injuries.

The societal impact of road accidents extends far beyond the loss of life. The OECD (2018) [3] estimates that the total social cost of road accidents in the European Union amounts to approximately 500 billion euros, or about 3% of the EU's GDP. In Croatia, road safety remains a considerable concern: statistics from the Ministry of the Interior indicate that in 2022 alone, 271 people lost their lives in road traffic accidents, 2,910 sustained severe injuries, and 10,419 suffered minor injuries, resulting in an estimated social loss of around 1.9 billion euros.

E. Macioszek et al. (Eds.): TSTP 2025, LNNS 1769, pp. 30–41, 2026.
https://doi.org/10.1007/978-3-032-14094-4_3

Despite these alarming figures, there have been some improvements. For example, Croatia achieved a 26% reduction in road fatalities between 2010 and 2018, demonstrating the potential effectiveness of targeted road safety interventions [4]. Nevertheless, the current number of fatalities, slightly less than 300 annually, and severe injuries, over 2,500 each year, underscore the ongoing need for robust, sustained efforts to further reduce road traffic casualties and meet the ambitious targets set by the European Union's "Vision Zero" strategy [5].

In addressing road safety issues, one of the key steps is to reduce the extent of speeding. Research consistently shows that road design characteristics, including features such as horizontal and vertical alignment, lane width, and roadside environment, significantly influence how fast drivers actually travel, often regardless of posted speed limits or enforcement measures. When there is a mismatch between the intended function of the road and its physical or visual cues, drivers may select speeds that do not align with the legal restrictions or the safe operation of the roadway. This persistent gap between desired and actual driving speeds presents a major challenge for effective speed management and underscores the urgent need for a deeper understanding of the factors shaping speed behavior on roads.

2 The Role of Speed in Traffic Accidents

When it comes to factors affecting the occurrence of traffic accidents, research shows that the causes are rarely attributable to a single factor; instead, they often result from a complex interplay of human, infrastructure, and vehicle-related elements. In Croatia, human factors such as speed, alcohol consumption, reckless driving, failure to use safety equipment, and distraction account for the majority of severe accidents, with speed contributing to 39% of cases [6]. Figure 1 shows the factors affecting severe traffic accidents in Croatia. The data presented in this paragraph related to Croatia is based on the national database of road traffic accidents which occurred from 2010 to 2019.

It can be noted that speed is an important factor affecting the occurrence of severe road traffic accidents. Alcohol is also a frequent concurrent factor, especially among younger drivers, and distraction, often related to mobile phone use, is implicated in up to 12% of severe accidents. Risky behaviors like not wearing seatbelts or helmets are particularly prevalent among younger road users and those involved in more serious crashes. The analysis also shows that vulnerable road users, including cyclists and motorcyclists, are significantly overrepresented in severe accidents, with improper use of safety equipment further increasing the risk of severe consequences [4].

Infrastructure and vehicle factors also play a critical role in shaping accident patterns and consequences. About 10% of severe accidents are linked to inadequate or missing road signage, with unclassified, national, and county roads being the most frequent locations, often where road design and speed limits are mismatched [4]. Vehicles involved in severe accidents frequently lack modern passive safety features, reflecting the risks posed by an aging vehicle fleet. Additionally, the quality and timeliness of post-crash medical intervention severely affect survival rates; studies suggest that reducing emergency response times can substantially lower fatalities.

One of the most widely-recognized factors regarding the occurrence of traffic accidents is speed. On an international scale, it is generally estimated that approximately

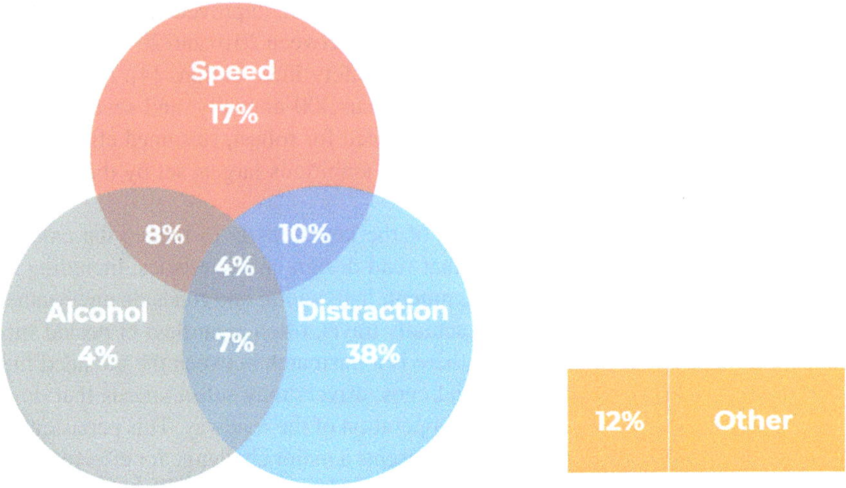

Fig. 1. Factors affecting the occurrence of severe traffic accidents in Croatia

one-third of fatal accidents are at least partially caused by excessive or inappropriate speed [7]. For example, data from Australia indicate that speed is responsible for around 41% of fatal road accidents and 24% of accidents resulting in severe injuries [8]. Similarly, on European roads, speed is identified as a causative factor in about 30% of all fatal accidents [9].

Croatia mirrors these global trends. Analyzing severe road traffic accidents in Croatia from 2010 to 2019, it was found that speed was a contributing factor in 18% of such cases, resulting in the deaths of 716 people and severe injuries to 4,770 others. Although the number of speed-related severe accidents decreased by 31% during the analyzed period, speed remains a dominant factor in both fatalities and serious injuries. More recent analyses reveal that speed was identified as a potential factor in 39% of severe road accidents, while it was the direct cause in approximately 17% of cases. Furthermore, the combination of traffic infrastructure and driver behavior is responsible for around 35% of severe accidents [4].

Empirical research further supports the considerable influence of speed on both the likelihood and severity of road traffic accidents. In particular, even a small increase in average speed has a disproportionate impact on accident risk: each 1% increase in average speed is associated with a 4% increase in the risk of fatal crashes and a 3% increase in the risk of crashes resulting in severe injury [1]. This relationship underlines why speed management has long been a central focus of global road safety policy. Nevertheless, given that speed remains a substantial contributor to serious accidents, there is still considerable potential for further improvements in this area, both at the EU level and within individual countries [10]. This is particularly urgent because high-speed accidents are far more likely to have severe consequences [11].

A more detailed examination of speed-related road traffic accidents in Croatia reveals several key patterns. Most commonly, speed-related accidents result in roadway departures (67%), followed by pedestrian collisions (21%) and roadside object collisions

(6%). Curves are the most frequent accident locations (53%), with straight segments accounting for 30% and intersections for 12% (mostly three- or four-leg intersections) [4].

These accidents are also seasonal, peaking in the summer months: 41% occur between June and September, with August and July being the most prevalent individual months (11% each). Roads rated as being of good quality are the site of the majority of speed-related accidents (96%), suggesting that perceived safety may encourage higher speeds. The most common road categories for speed-related accidents are uncategorized roads (47%) and state roads (29%), followed by county roads (15%), highways (5%), and local roads (4%). Additionally, these accidents tend to occur during the day, on unlit roads, and most frequently involve passenger cars and drivers aged 19–34 [4].

These findings highlight the importance of understanding when, where, and why speeding occurs in order to develop more targeted and effective countermeasures. Despite substantial progress in speed control measures and policy interventions over recent decades, the persistent prevalence of speed in serious accidents highlights the untapped potential for further advances. Numerous studies have, therefore, focused on understanding the relationship between vehicle speed, road infrastructure, and various external factors, including geometric road parameters, traffic characteristics, and driver behavior.

3 The Impact of Inadequate Road Design on Speed and Safety

A persistent challenge in road safety is the gap between operating speed, the speed most drivers actually travel at under normal conditions, design speed, the maximum speed for which a road was engineered, and the posted speed limit. In many cases, the physical and visual characteristics of a road encourage higher speeds than what is legally allowed or safe for the specific context. Elements such as wide lanes, long sight distances, gentle curves, or the absence of traffic-calming features can signal to drivers that it is "safe" to travel faster, even when surrounding land use or traffic conditions do not support such speeds. This mismatch creates a situation where speed limits are routinely disregarded, and the effectiveness of speed management measures is undermined.

When it comes to measuring speeding, among all types of speed metrics, operating speed stands out as particularly important due to its frequent deviation from posted speed limits. Operating speed, or the 85th percentile speed, is typically defined as the speed at which drivers travel under free-flow conditions and in the absence of speed enforcement. Research consistently shows that operating speeds are often higher than speed limits. For instance, a study conducted in Bosnia and Herzegovina found that over 90% of vehicles exceeded the posted speed limit, with the average operating speed surpassing the limit by around 15 km/h [12]. Similar patterns are observed in Croatia, where data from all national road traffic counters on state roads for the entire year of 2021 indicate that the average difference between operating speed and the speed limit was substantial, ranging from 26 km/h over the limit in 40 km/h zones to 8 km/h in 90 km/h zones. Notably, the deviation is inversely proportional to the speed limit; lower speed limits see greater differences between actual speeds and posted limits [13].

While design speed and calculated (or legal) speed are determined by road planners, authorities, and regulations, operating speed depends on a combination of driver behavior, vehicle characteristics, and road environment. Across Europe, it is estimated that 40–60% of drivers do not adhere to speed limits. The rate of non-compliance varies by road type: 28% of European drivers admit to speeding on motorways, compared to 19% on state roads, 13% on local roads, and just 7% in built-up urban areas. Drivers tend to respect speed limits they perceive as reasonable or appropriate for the road's characteristics; if not, they are more likely to ignore them, which can undermine the entire speed management system [14].

To better understand speeding behavior, comprehensive analyses have been carried out on motorways, state roads, and urban roads in Croatia. This research was conducted on a set of representative road sections, and only free-flow conditions were taken into consideration. All road segments which were analyzed were chosen based on the absence of speed enforcement infrastructure, such as cameras or permanent speed measurement devices. Additionally, only segments with relatively uniform geometric characteristics, such as the absence of curves, intersections, or other disruptive elements, were included. Additionally, the locations had to be representative of different geographical areas of the country.

Highway road sections had to have free-flow conditions, and a total of 8.5 thousand vehicles were observed in such conditions on several road sections of three different highways. Highway data was analyzed on a sample of 22 cross-sections, state road data was analyzed on a sample of 118 different cross-sections, while in urban areas, 61 cross-sections were chosen [4].

It is vital to note that the default speed limits (unless stated otherwise) in Croatia are 50 km/h in built-up areas, 90 km/h outside of them and 130 km/h on motorways.

On Croatian motorways, measured speeds consistently exceeded legal limits, with a median speed of 127 km/h, an average speed of 125 km/h, and an 85th percentile speed of 145 km/h, despite varying posted limits. The standard deviation of speeds was relatively high at 21.5 km/h, and more than half of drivers were recorded exceeding the speed limit at the surveyed locations [4].

On state roads, an analysis across 118 locations found the median 85th percentile speed to be 90 km/h (average 92 km/h, 85th percentile of the dataset 109 km/h, standard deviation 14.2 km/h). Notably, the difference between the 85th percentile speed and the speed limit was found to decrease as the speed limit increased, indicating that drivers are more likely to exceed the limit in lower-speed zones. Speeding on state roads was found to be marginally more common at night, at locations with road shoulders or pedestrian footpaths, and among passenger cars compared to heavy goods vehicles. Interestingly, higher volumes of daily or summer traffic correlated with a lower proportion of speeding drivers.

Urban environments present a particularly acute issue. Measurements within urban areas showed a median and average 85th percentile speed of 82 km/h, with a standard deviation of 8.8 km/h, well above typical urban speed limits. More granular data from selected urban locations in Zagreb revealed that both the median and 85th percentile speeds frequently exceeded the speed limit, with the share of speeding drivers ranging from 33% to as high as 80%. On average, 60% of drivers at these sites were recorded as speeding. Such widespread non-compliance is especially troubling given the increased risks to vulnerable road users in urban settings [4].

These findings collectively underscore that operating speed often significantly exceeds legal speed limits across various road environments, especially in lower-speed and urban zones. This persistent behavior highlights the complex interplay between infrastructure design, driver perception, and enforcement, suggesting that substantial room remains for targeted interventions to reduce speeding and improve overall road safety. It should be noted that speeding tickets in Croatia can be considered as not high enough, especially outside of built-up areas. This leads to a smaller fear of being fined, which leads to speeding. This further underscores the importance of eliminating the potential for speeding through thoughtful design from the design phase.

Traditionally, excessive speed has been attributed primarily to driver behavior. However, a growing body of research and practice suggests that road design plays a notable role in shaping driving speeds. Roads that are over-designed for their intended context, such as urban streets built to highway standards, can inadvertently encourage drivers to exceed safe speeds. In such cases, blaming drivers alone oversimplifies the issue and overlooks the responsibility of road planners, designers, and managers. If the design speed or visual cues provided by the road environment suggest higher speeds than the posted limit, it is predictable that many drivers will follow those cues rather than the signage. Therefore, the interplay between human factors and engineering design must be considered when addressing speeding, rather than focusing solely on enforcement and education.

Several Croatian examples illustrate the real-world implications of these issues. On the Turanj bypass, Vodice bypass, and sections of the D1 State Road in the Krapina-Zagorje County, there have been ongoing concerns regarding the relationship between road design and driver speed choices. On these bypasses, the roadway geometry, such as long, straight sections and broad cross-sections, encourages operating speeds far above the posted limits. Despite the intended function of these roads (often as regional or local connectors), their physical design is more consistent with roads intended for much higher speeds, lacking sufficient measures to induce drivers to comply with speed limits or adjust their speed to match local conditions.

These cases underscore the need to align road design with the intended function and surrounding environment. Without such alignment, efforts to manage speeds through enforcement or signage alone are likely to be less effective, as drivers respond more strongly to environmental cues than to posted numbers. This further highlights that addressing speeding is not solely a matter of driver behavior, but a systemic challenge that involves the road design process, infrastructure management, and policy.

Research consistently shows that the geometric and environmental characteristics of road infrastructure are among the central factors influencing vehicle operating speeds. One of the primary findings is that horizontal alignment, especially the radius of horizontal curves, is statistically the most pronounced variable affecting operating speeds on all road types. When the radius of a horizontal curve is smaller than approximately 300 m, drivers drastically reduce their speed; however, for radii greater than 400 m, the impact on operating speed becomes negligible. Other noteworthy factors in curves include the curve length and deflection angle, both of which can also influence how drivers adjust their speed, longer curves and smaller deflection angles generally result in higher operating speeds [13].

For straight road segments, or tangents, two variables stand out: the length of the tangent and the posted speed limit. Drivers tend to accelerate and reach their desired speeds on longer straight segments, while shorter ones do not allow enough time for such acceleration. The posted speed limit is especially important on tangents and in vertical curves, where additional factors like lane width and longitudinal slope (gradient) can also have a measurable effect. Wider lanes are often associated with higher operating speeds, and road segments with smooth or gentle gradients may encourage faster driving. On the other hand, greater pavement roughness and unevenness tend to reduce operating speeds, likely due to reduced comfort and perceived safety [13].

Vertical alignment is another pivotal design element. On downhill segments (descents), the initial speed, posted speed limit, and longitudinal gradient are all statistically significant factors. Generally, a downhill gradient up to 3% increases speed, but steeper slopes beyond this point tend to decrease speeds due to safety concerns. On convex and concave vertical curves, sight distance is also a vital factor; reduced visibility causes drivers to slow down. Research has shown that the combined effect of horizontal and vertical curves can strongly influence speed, with the combination of a downhill and a curve producing the greatest reduction in speed [13].

Other environmental and cross-sectional elements also play a role. The presence of shoulders (especially paved shoulders), type of median, number and width of lanes, and even roadside features like parking, pedestrian traffic, and access points can influence speeds. For example, more access points (driveways, minor roads), pedestrian crossings, and visible roadside activity tend to lower operating speeds, especially in urban environments. In contrast, open landscapes with fewer visual cues or obstacles generally lead to higher speeds [13].

Finally, the design of consecutive geometric elements (such as sequences of curves and tangents) can affect the consistency of operating speeds. When the speed differential between consecutive elements (e.g., between a straight and a curve) exceeds about 10–20 km/h, the design is considered less consistent and may lead to sudden speed changes, increasing the risk of accidents. Good design practice seeks to minimize abrupt speed changes by ensuring that transitions are gradual and predictable for drivers [13].

In summary, the most influential road design factors affecting speeding are:

- Radius, length, and deflection angle of horizontal curves,
- Length of straight segments (tangents),
- Posted speed limit,
- Lane and shoulder width,
- Longitudinal and vertical alignment (gradients and curves),
- Pavement quality and roughness,
- Presence of medians, access points, and pedestrian activity,
- Consistency and sequence of geometric elements.

These factors interact to shape drivers' perceptions and behaviors, often more strongly than posted speed limits or enforcement measures alone. Thus, understanding and optimizing these design elements is crucial for effective speed management and road safety.

The aforementioned Turanj Bypass can be used as an example of inadequate or mismatched road infrastructure. A series of post-opening audits revealed several infrastructural deficiencies directly impacting traffic safety. For example, the field inspections and speed measurements showed that a large proportion of drivers travel well above the posted speed limits. During peak tourist weeks, almost half of all vehicles were found to be exceeding 90 km/h, even though the speed limit was set at 80 km/h; about 40% were observed driving over 100 km/h. Similar behaviors were recorded outside the tourist season. This discrepancy is partly attributed to the fact that the physical and geometric design elements (such as wide lanes, gentle curves, and sight distances) unintentionally encourage higher speeds, making the posted limits appear unreasonably low to drivers.

Another recurring issue is the lack of clear and consistent traffic signage and road markings, which can cause confusion, particularly in transition zones (such as the change from two to one carriageway), intersections, and curves. For instance, several horizontal curves along the bypass were found to be insufficiently marked, reducing driver awareness and increasing the risk of run-off-road and loss-of-control accidents. The audit also highlighted numerous cases of non-standard or incomplete road safety equipment, such as improperly installed or missing guardrails, insufficient end treatments, and inadequate transitions between different types of barriers. At several locations, abrupt changes in barrier types or unprotected barrier ends were identified as potential hazards for errant vehicles, with a real risk of severe consequences in the event of impact.

Moreover, the design often failed to anticipate the real-world behavior of drivers. For example, the lack of adequate safe stopping zones (emergency lay-bys) along the route was noted as a serious shortcoming, especially considering the high operating speeds and substantial traffic volumes during the tourist season. In case of a breakdown or emergency, vehicles have limited safe options for stopping, which could lead to dangerous situations. Similarly, visibility issues at certain intersections and the lack of dedicated lanes or warning systems for complex traffic movements further compounded safety risks.

4 Engineering Solutions: Implementing Self-Explaining and Forgiving Road Principles

Several options exist which can be used to adequately inform drivers about the speed at which they should be driving. The concept of self-explaining roads (SER) is rooted in the idea that the road environment itself should clearly communicate its intended use to drivers, prompting correct and safe behavior without relying primarily on signs or enforcement. First introduced in the Netherlands in the 1990s by Theeuwes and Godthelp, SER draws on psychological principles: when a road's visual, geometric, and environmental cues are consistent with its function and expected speed, drivers intuitively adjust their speed and behavior accordingly [15]. Key SER features include consistent use of lane widths, markings, roadside development, landscaping, and intersection treatments that "signal" to users the appropriate way to behave. These measures should be implemented in the design phase. If not implemented in the design phase, other measures are also required, such as traffic calming measures, enforcement measures, etc.

For example, local residential streets are designed to appear narrow, include traffic-calming elements, and have active frontages and pedestrian presence, all of which indicate low speeds. In contrast, arterial roads have wider lanes, clearer separation from pedestrians, and longer sight distances, naturally supporting higher speeds. Empirical studies have shown that self-explaining designs reduce speeding and crash rates. A notable field experiment in the Netherlands demonstrated that redesigning a road to match SER principles led to considerable decreases in both mean speed and speed variance among drivers [16].

European best practice, as seen in Dutch, Danish, and Swedish policy, increasingly mandates self-explaining principles, moving beyond enforcement toward designing for behavior. SER is now central to the European Union's Safe System approach and the UN's Decade of Action for Road Safety.

While SER focuses on preventing errors, the concept of forgiving roads (or forgiving infrastructure) recognizes that human mistakes are inevitable and seeks to minimize the severity of crashes when they occur. The forgiving road philosophy is a core tenet of the Safe System approach, which holds that no human loss of life is acceptable and that the road system must accommodate human fallibility [17].

Forgiving roads feature:

- Wide and obstacle-free shoulders and clear zones, allowing vehicles to recover safely if they leave the carriageway.
- Energy-absorbing guardrails and crash barriers to minimize injury if a collision occurs.
- Rumble strips and tactile edge lines to alert drivers before they drift off the road.
- Gradual side slopes and non-rigid objects along the roadside, reducing crash severity.
- Well-designed intersections with features such as roundabouts or protected turn lanes to reduce the potential for severe conflicts.

Meta-analyses and international crash data show that implementing forgiving design elements, particularly on rural and higher-speed roads, can substantially reduce the number of fatal and serious injury crashes [18].

Many national and regional guidelines, such as Austroads in Australia, the Swedish Vision Zero program, and Dutch road categorization systems, now require roads to be both self-explaining and forgiving [19]. In Croatia and Central Europe, recent revisions to road design standards also incorporate these concepts, as reflected in your document [13].

These principles are particularly important in the context of speed management. If a road appears safe for high speeds (wide, straight, few obstacles), drivers will speed regardless of posted limits, unless the design is modified. Likewise, forgiving road features help minimize the severity of crashes resulting from speed-related errors or loss of control, a pivotal consideration given the widespread mismatch between operating speeds and posted limits observed in Croatia and across Europe.

Self-explaining and forgiving roads work in tandem: the former encourages correct user behavior through intuitive design, while the latter mitigates the consequences of inevitable mistakes. The integration of both approaches is crucial for achieving substantial, sustainable reductions in road fatalities and serious injuries. Modern road safety policy at the EU and global level treats these philosophies as central pillars for future road system development.

Addressing these issues requires a multi-faceted approach that goes beyond mere compliance with minimum technical standards. One crucial recommendation is the rigorous application of comprehensive road safety audits (RSA) at all stages of road design, construction, and operation, not just post-opening. The audits should focus on identifying and rectifying elements that could mislead drivers or encourage unsafe behaviors. For example, on the Turanj Bypass, it was recommended to use double solid lines with a vibration effect to clearly indicate no-passing zones and to reinforce the ban on overtaking, especially where geometric elements could otherwise invite risky maneuvers.

Implementation of intelligent speed enforcement systems, such as average speed cameras (Section Control), was also advised to manage operating speeds more effectively over the entire corridor, rather than relying solely on spot enforcement. This can help align actual driving behavior with intended safety consequences, especially in areas where geometric road elements unintentionally support higher speeds. Where feasible, consideration should also be given to adjusting speed limits so that they better reflect the road's characteristics and driver expectations, provided this does not compromise safety.

A further set of solutions involves improving the clarity and intuitiveness of road markings and signage, particularly at curves, intersections, and transition zones. Enhanced use of visual and tactile cues, such as warning strips, well-placed signs, and prominent directional arrows, can help guide driver behavior, making the infrastructure more "self-explaining." Where there are frequent queues or complex movements (as at T-junctions or in areas with poor visibility), the deployment of variable message signs and early warning systems can increase driver attention and reduce the risk of rear-end or side-impact collisions.

In addition, upgrading and standardizing safety barriers and their transitions is essential to prevent barrier ends and interfaces from becoming hazards themselves. Certified transitions and energy-absorbing terminals should be systematically installed at all critical points. Attention should also be given to the provision of safe stopping areas, particularly on long stretches without frequent exits or shoulders, to accommodate emergencies without compromising the safety of other road users.

Collectively, these measures not only address the immediate deficiencies identified but also move the network closer to the principles of self-explaining and forgiving roads, where infrastructure intuitively supports safe behavior and mitigates the consequences of human error.

5 Conclusions

This analysis highlights the complex and multifaceted relationship between road design, speed, and road safety consequences. While significant progress has been made in reducing road fatalities and serious injuries, both globally and in Croatia, the persistent prevalence of speed-related accidents demonstrates that much work remains. The evidence clearly shows that inadequate or mismatched infrastructure, particularly the gap between operating, design, and posted speeds, continues to undermine the effectiveness of speed management strategies and contributes to high-risk scenarios, especially on regional connectors such as the Turanj and Vodice bypasses and the D1 corridor.

Rather than placing sole responsibility on drivers, this research underlines the need for a systemic approach that considers the powerful influence of infrastructure design on road user behavior. The analysis demonstrates that road geometry, cross-sectional elements, signage, and the overall road environment strongly shape operating speeds, often more so than enforcement or education alone. Consequently, interventions must go beyond minimum standards and traditional measures to address the underlying causes of unsafe speed selection.

International experience and best practice, especially the adoption of self-explaining and forgiving road principles, offer a clear path forward. By designing roads that intuitively guide users toward appropriate behaviors and minimize the consequences of inevitable human error, policymakers and engineers can achieve sustainable reductions in fatalities and serious injuries. For Croatia, this means rigorously auditing infrastructure, aligning design with function, deploying intelligent speed management, and systematically integrating these modern concepts into both new projects and upgrades of the existing network.

Further research is recommended to compare driver behavior under varying conditions. Specifically, future studies should include detailed comparisons of similar road segments while accounting for differences in weather conditions, levels of enforcement, driver demographics, and the inclusion of appropriate control sites.

In summary, the challenge of managing speed and enhancing road safety cannot be solved by targeting drivers alone. Achieving "Vision Zero" and similar ambitious targets will require a holistic, evidence-based approach in which safe infrastructure plays a central role, creating a road system that both anticipates human behavior and protects all users.

References

1. World Health Organization: Global status report on road safety 2018. World Health Organization (2018)
2. Ministarstvo unutarnjih poslova Republike Hrvatske: Nacionalni plan sigurnosti cestovnog prometa Republike Hrvatske za razdoblje od 2021. do 2030. Ministarstvo unutarnjih poslova Republike Hrvatske, Zagreb (2021)
3. ITF: Road Safety Annual Report 2018. OECD Publishing, Paris (2018)
4. Šoštarić, M., Ševrović, M., Jakovljević, M., Gruhonjić, D., Štrkalj, Z.: Speed as a key factor of road traffic safety and potential methods for mitigating its impact. In: Macioszek, E., Granà, A., Sierpiński, G. (eds.) Road Traffic Research, Analysis, and Modeling. TSTP 2024. Lecture Notes in Networks and Systems, vol. 1315. Springer, Heidelberg (2025)
5. Ministarstvo unutarnjih poslova Republike Hrvatske: Bilten o sigurnosti cestovnog prometa za 2023. godinu. Ministarstvo unutarnjih poslova Republike Hrvatske, Zagreb (2024)
6. Department of Transport Planning: Analysis of Critical Factors Influencing the Occurrence of Traffic Accidents. University of Zagreb, Zagreb, Faculty of Transport and Traffic Sciences (2020)
7. SWOV: Speed and speed management, SWOV Fact sheet, July 2021. SWOV Institute for Road Safety Research, The Hague (2021)
8. Transport for NSW: Roads and Waterways. Transport for NSW, Sydney (2010)

9. European Automobile Manufacturers Association: Road Safety: safe vehicles, safe drivers, safe roads. European Automobile Manufacturers Association (2019)
10. European Transport Safety Council: Reducing Speeding in Europe, PIN Flash Report 36. European Transport Safety Council (2019)
11. Eluru, N., Bhat, C.R.: A joint econometric analysis of seat belt use and crash-related injury severity. Acc. Anal. Preven. **39**(5), 1037–1049. Elsevier (2007)
12. Subotić, M., Stepanović, N., Tubić, V., Softić, E., Bouraima, M.B.: Models of Analysis of Credible Deviation from Speed Limits on Two-Lane Roads of Bosnia and Herzegovina. Complexity, vol. 2022. Wiley (2022)
13. Vertlberg, J.L., Švajda, M., Jakovljević, M., Ševrović, M.: Operating vehicles' speed prediction models. Promet – Traffic Transp. **36**(3), 383–398 (2024)
14. Bonić A.M., Salkanović, M.: Transport and logistics analysis of operating speed deviations from limit values on two-lane roads. In: The Seventh International Conference Transport and Logistics, pp. 133–138 (2021)
15. Theeuwes, J., Godthelp, H.: Self-explaining roads. Saf. Sci. **19**(2–3), 217–225. Elsevier (1995)
16. Aarts, L., van Schagen, I.: Driving speed and the risk of road crashes: a review. Ac. Anal. Prevent. **38**(2), 215–224. Elsevier (2006)
17. OECD/ITF: Towards Zero: Ambitious Road Safety Targets and the Safe System Approach. International Transport Forum, Paris (2008)
18. Elvik, R., Vaa, T., Hoye, A., Sorensen, M.: The Handbook of Road Safety Measures, 2nd edn. Emerald Group Publishing, Leeds (2009)
19. Austroads: Guide to Road Design Part 1: Objectives of Road Design, Austroads Ltd, Sydney (2021)

Hourly Distribution of the Probability of a Fire Accident in a Urban Road Tunnel

Aleksander Król(✉), Małgorzata Król, and Krystian Szewczyński

Silesian University of Technology, Gliwice, Poland
{aleksander.krol,malgorzata.krol,krystian.szewczynski}@polsl.pl

Abstract. Urban road tunnels are a reasonable remedy for the inconvenience due to the congested road traffic. However, they bring specific threats, especially those related with the possibility of fire outbreak. The work is a case study for an urban road tunnel, and uses detailed data on traffic intensity and its generic structure. It was shown that although tunnel fire accidents are very rare events, a single fire can be expected in a relatively short time perspective, up to 10 years. The probability of a catastrophic fire is only serious in longer time horizons (about 50 years). The next important finding is that the hourly distribution of the fire accident probability is uneven, it is significantly higher at daytime, reaching the maximum value at 3 pm. This probability on weekends is significantly lower than for workdays. Such quantitative findings may serve as a basis for decisions on temporary restrictions on heavy vehicle traffic.

Keywords: urban road tunnel · tunnel fire · fire accident likelihood

1 Introduction

Road tunnels are an important component of transportation infrastructure to overcome terrain obstacles while reducing travel distance and time. Other advantages of tunnels include insensitivity to weather conditions, which is also associated with a reduction in the daily amplitude of air temperature changes.

For cities, road tunnels are an effective alternative to ring roads. They run under city centers, thus eliminating the need to plan roads around a highly urbanized city center and avoiding the difficulties of purchasing land for road construction. Urban road tunnels relieve the burden on the surface transportation network and allow for a more convenient arrangement of urban space. They are a frequent solution for reducing urban traffic, associated noise, and traffic pollution.

Road accidents in tunnels are rarer than at other sections of the transportation network because a tunnel itself calms down the traffic. In the tunnel there are often speed limits, no intersections, no pedestrian traffic, and the impact of the weather is negligible [1, 2]. On the other hand, the peculiarities of the tunnel (limited space, difficult access) bring with them new risks, which in extreme cases can lead to catastrophic events [3]. These catastrophic events are most often related to fires.

E. Macioszek et al. (Eds.): TSTP 2025, LNNS 1769, pp. 42–54, 2026.
https://doi.org/10.1007/978-3-032-14094-4_4

Fire accidents can significantly disrupt the operation of urban roads, and this concerns not only tunnels. But significant disruption to urban traffic affecting the whole city is only an addition to events occurring in the tunnel (risk to people and the tunnel structure). Extensive research is needed to indicate a methodology for predicting these disruptions and to approximate the process of restoring traffic flow throughout the city [4].

As stated above, the most important hazard associated with the use of road tunnels is the possibility of fire. In this case, in addition to people in the immediate vicinity of a fire event, all tunnel users are at risk. This is because the main factor that endangers people is the rapidly spreading smoke. The smoke generated during a vehicle fire is toxic and additionally limits visibility and confuses people [3]. This factor can be even more dangerous in urban road tunnels due to their greater congestion with vehicles. The high temperatures associated with vehicle fires endanger people, as well as tunnel structures [5]. Numerous studies are being carried out on the appropriate design of tunnels in terms of fire resistance [6–8]. Such research is of great importance to ensure the stability of the tunnel vault structure during a fire. The safety of people in a road tunnel where a fire is developing is also often analyzed.

The likelihood of loss of life for people in a tunnel where a fire is developing is examined [9]. Such analyses take into account where people are in the tunnel in relation to the location of emergency exits, the location of the fire development, or with regard to different traffic conditions [10, 11]. The issues related to the safe use of road tunnels and the risks associated with the development of fire in tunnels are strongly linked to the problem of detecting fire at the earliest possible stage. Nowadays, they are focused mainly on improving image analysis tools [12, 13].

Szewczyński and others, collecting data from many sources, estimated the probability of a road tunnel fire accident. They also proposed the exponential distribution for fire powers and assessed the average fire power for non-trucks and trucks [14].

The main contribution of the present work is to analyze the daily probability of a fire accident in an urban road tunnel hour by hour, taking into account detailed data on traffic intensity and its generic structure. The novelty lies in fact that in deep analyzes allowed to indicate time periods of especially high risk of a fire accident.

The rest of the paper is organized as follows: first the examined tunnel is described and the traffic characteristic is provided, then general considerations on the probability of a fire accident are posted. In the Results section, the ideas introduced are used to estimate the probability of a tunnel fire in different time perspectives. The Conclusions section highlights the most important findings.

2 The Examined Tunnel

2.1 Description of the Tunnel and Its Sensor

The examined tunnel is an urban tunnel located in Gdansk (Poland) under one of the branches of the Dead Vistula River. The total length of the tunnel is 1378 m and it consists of two tubes for both directions of traffic. There are two traffic lanes in each tube. The minimum depth under the river bed is about 8 m. The tunnel runs from southeast (SE portal) to northwest (NW portal). The north tube is referred to as T1 and runs from SE to NW, and the south tube is referred to as T2 and runs from NW to SE (Fig. 1).

Traffic parameters were recorded by MC3224 inductive loop detectors manufactured by SWARCO TRAFFIC SYSTEMS. These were classification detectors operating with two induction loops per lane. The detectors provided data such as vehicle class, speed (up to 300 km/h), length, direction of travel and distance between vehicles. Classification was carried out according to German standards; vehicles were classified as cars, trucks, large goods vehicles (HGVs), HGVs with trailers and buses (TLS, 2012). The quality of detection was not affected by weather conditions, such as wet roadway surfaces [15].

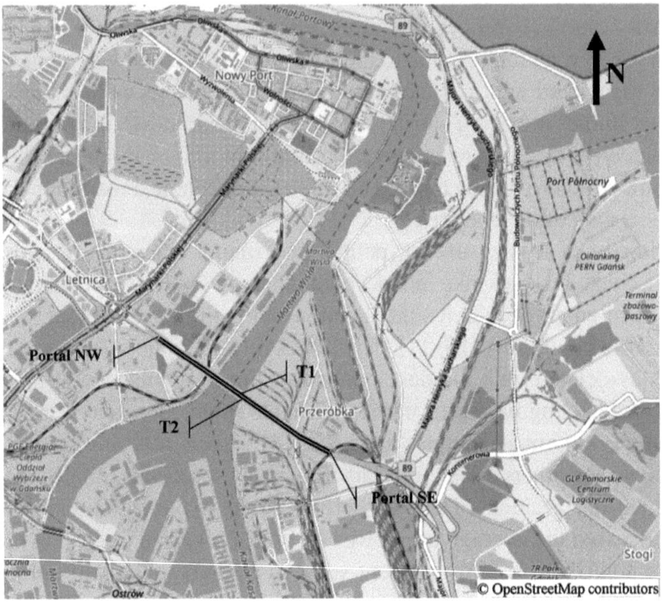

Fig. 1. The general layout of the tunnel

2.2 Traffic Characteristics

Since the fire risk depends on the traffic intensity and its generic structure, data on the traffic properties are crucial to a accurate assessment of the probability of the occurrence of a fire accident. Data from tunnel sensors shared by the management of the tunnel under Martwa Wisła in Gdańsk were used in the work.

The data of road traffic cover two periods: a week from 15.08.2021 to 22.08.2021 and a week from 27.09.2021 to 3.10.2021. They include traffic intensity for each lane, average speed for each lane, shares of passenger cars, light trucks, trucks, and buses. Data were recorded every 10 s for both tunnel tubes separately. So, the whole dataset contains 241 920 records.

Figures 2 and 3 presents the daily distribution of traffic intensity over both examined periods for both tunnel tubes separately.

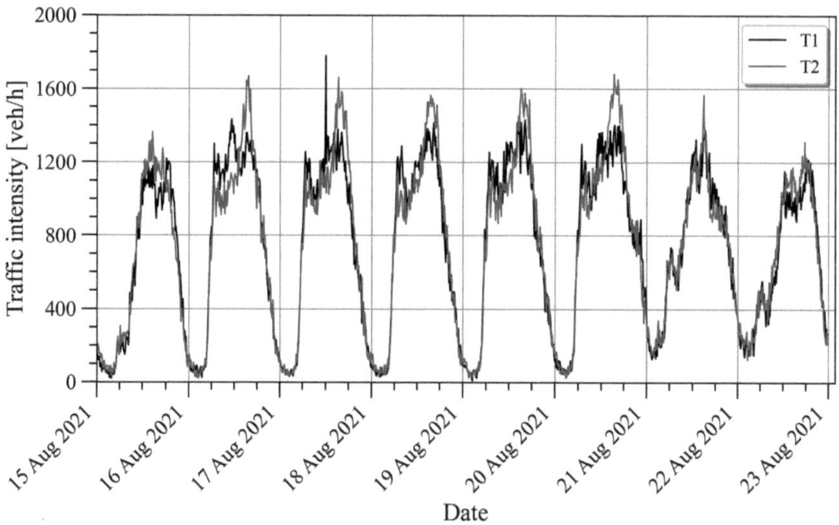

Fig. 2. Traffic intensity for period 15.08.2021 to 22.08.2021

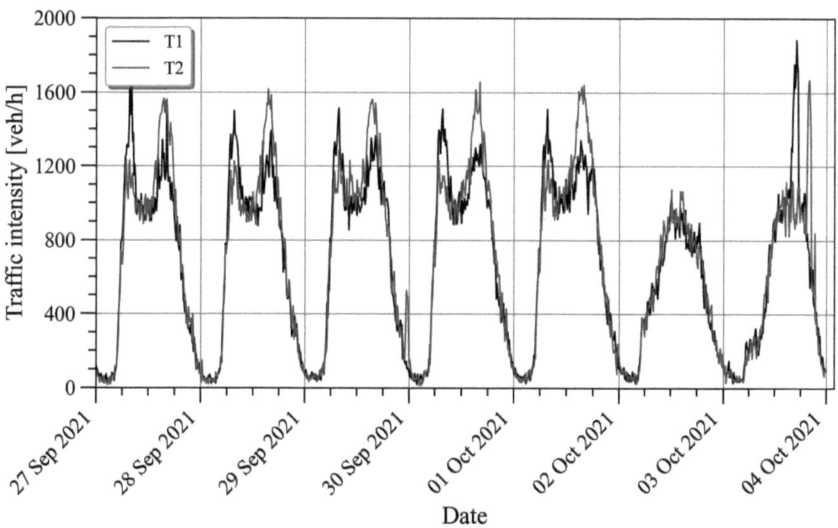

Fig. 3. Traffic intensity for period 27.09.2021 to 3.10.2021

In the second period, in the autumn, there are two peaks visible each day excluding weekends, which correspond to the rush hours. These peaks are not so clear for the first period, in the summer. Two unusual peaks can easily be seen on Oct 4th at around 5 pm for the T1 tube and around 9 pm for the T2 tube. They were caused by the mass commuting to and returning from a football event that had just been held at the stadium located on the road that runs through the tunnel, less than a kilometer from the NW portal. Figures 4 and 5 present the averaged data on traffic intensity for both periods and both tunnel tubes for workdays and weekends separately.

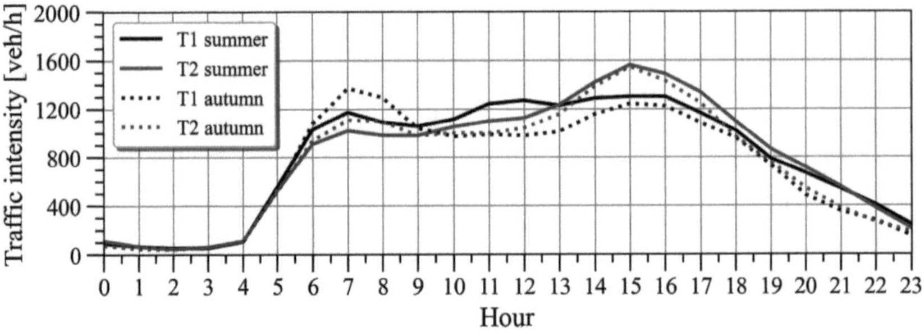

Fig. 4. Averaged hourly distribution of traffic intensity for workdays

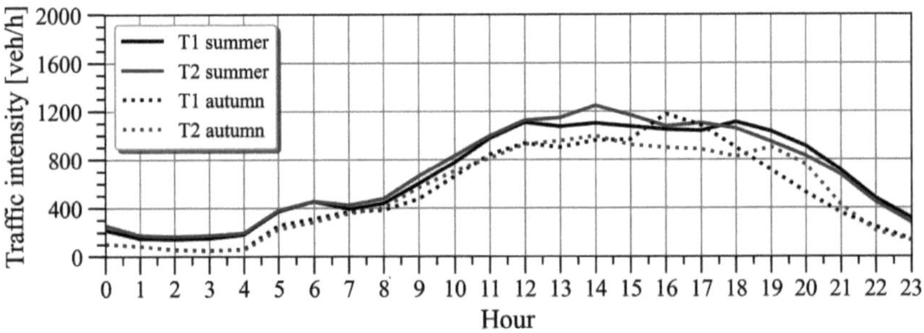

Fig. 5. Averaged hourly distribution of traffic intensity for weekends

As can be seen, according to expectations, traffic on weekends is much less intense than on workdays. Traffic patterns are the same, as revealed in Figs. 2 and 3, except that all extraordinary peaks are smoothed. The average traffic intensity in the examined period is 710 veh/h.

Figures 6 and 7 show the shares of passenger cars, HGV and buses. Since the hourly distributions of share of different vehicles' types for both tubes and both periods have appeared very similar, they are presented only in two figures: for workdays and weekends separately.

One can observe that passenger cars prevail. This is especially true on weekends, when the share of other types is very small. The share of buses is small all the time, it doesn't exceed 1%. On workdays, the HGV share reaches almost 20%, but it happens at night, when the total traffic intensity is very low, so the absolute number of HGVs is not too high.

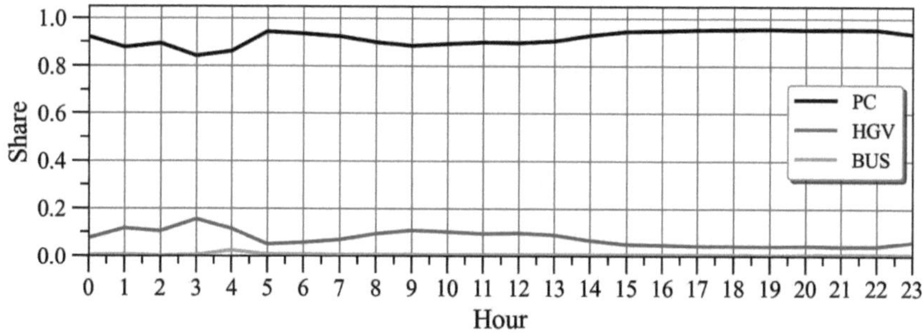

Fig. 6. Hourly distribution of share of different vehicles' types for workdays

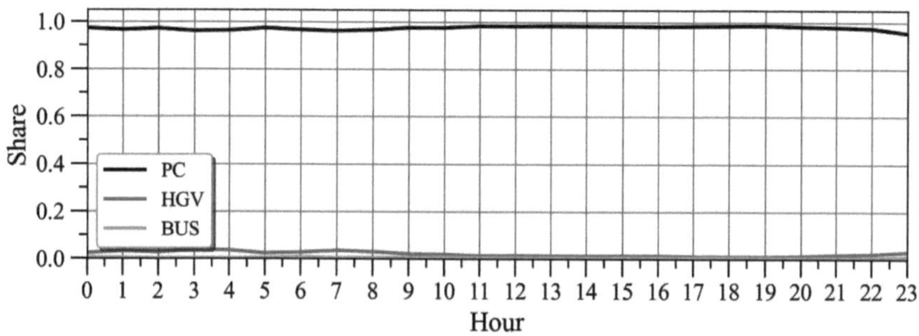

Fig. 7. Hourly distribution of share of different vehicles' types for weekends

The average share of HGV is about 10% during daytime, and is slightly lower at rush hours (simply there are many passenger cars in these moments). During the rest of the time the HGV share is about 5%. The average value for the considered period is 6.3%.

3 Probability of a Fire Accident in an Urban Road Tunnel

The average frequency of fire accidents in road tunnels (*FFA*) was estimated to be 8 accidents per billion vehicle kilometers travelled in tunnels [14]. However, it was observed that it is much higher for heavy goods vehicles (HGV). According to data from Austria, which are the only providing this explicitly, the appropriate multiplier for non HGV vehicles should be adopted as $m_{\sim HGV} = 0.646$, and for HGV as $m_{HGV} = 3.846$ [16]. Additionally, this value depends on the type of the tunnel, and is a bit lower for urban unidirectional tunnel (such as the examined tunnel). Using the data from France another multiplier can be introduced for this type of tunnels: $m_{UU} = 0.818$ [17].

Fires in tunnels are rare events and occur independently of each other, so their occurrence is governed by the Bernoulli distribution. A single attempt corresponds to every time a vehicle passes through a tunnel, a successful outcome occurs if a fire breaks out. If the number of trials (n) rises to infinity and the probability of a single success (p) diminishes to zero, such that the product is constant, this distribution can be approximated by the Poisson distribution. Practically, this becomes true for $n > 20$ and $p < 0.2$. Hence, the probability that k tunnel fires occur in a series of n tunnel travels can be expressed as follows:

$$P(k) = \frac{(np)^k}{k!} \mathrm{e}^{-np} \tag{1}$$

The number of attempts depends on the traffic intensity (T_I) and the period considered (c). If both quantities are defined by the same time unit (for example, an hour), and the tunnel length is L km, the product np can be expressed as (FFA_x denotes the basic value of the frequency of fire accidents multiplied by the appropriate modifiers):

$$np = L\frac{FFA_x}{10^9} cT_I \tag{2}$$

The np product expresses the expected number of tunnel fires in the examined period. For vehicles other than trucks, for the tunnel considered (which is a urban unidirectional) the actual value of frequency of fire accidents equals $FFA_x = FFA \cdot m_{UU} \cdot m_{\sim HGV}$, and for trucks $FFA_x = FFA \bullet m_{UU} \cdot m_{HGV}$. Knowing the share of HGV (x_{HGV}), one can determine the expected number of fire accidents for non-trucks and trucks as follows:

$$np_{\sim HGV} = L\frac{FFA \cdot m_{uu} \cdot m_{\sim HGV}}{10^9} cT_I(1 - x_{HGV}) \tag{3}$$

$$np_{HGV} = L\frac{FFA \cdot m_{uu} \cdot m_{HGV}}{10^9} cT_I x_{HGV} \tag{4}$$

It was observed that a fire of a single passenger car commonly doesn't pose commonly a very serious threat to majority of the people inside a road tunnel, because the tunnel systems are expected to cope in such a situation, and the zone of non-tenable conditions should be limited to the close vicinity of the fire accident [18]. However, it is possible that a catastrophic fire of very high power outbreaks. Such an event must involve an HGV vehicle [18] and the fire power should exceed 20 MW [19]. The distribution of the probability of fire power is assumed to be exponential with an average $\lambda = 3.96$ MW [14]. Thus, for catastrophic fires, the np_{HGV} product introduced by Eq. (4) must be multiplied by the probability of a fire stronger than $fp_{cat} = 20$ MW. Thus, the expected number of catastrophic fires is:

$$np_{HGV,cat} = np_{HGV} e^{-\frac{fp_{cat}}{\lambda}} \tag{5}$$

4 Results and Discussion

In the first phase of the research some general predictions on expected number of fires and probability of fire in long term horizons were made using formulas introduced above. This is because, happily, fire accidents in road tunnels are rare events, the probability of such an accident takes significant values only in longer time horizons. Data from both periods examined were treated as representative for the tunnel, and it was assumed that traffic characteristics will be almost the same in the future. Since the share of buses is negligible, the calculation was carried out noting only the share of passenger cars and HGVs.

Figure 8 shows the probabilities of the occurrence of different number of tunnel fires in selected time perspectives.

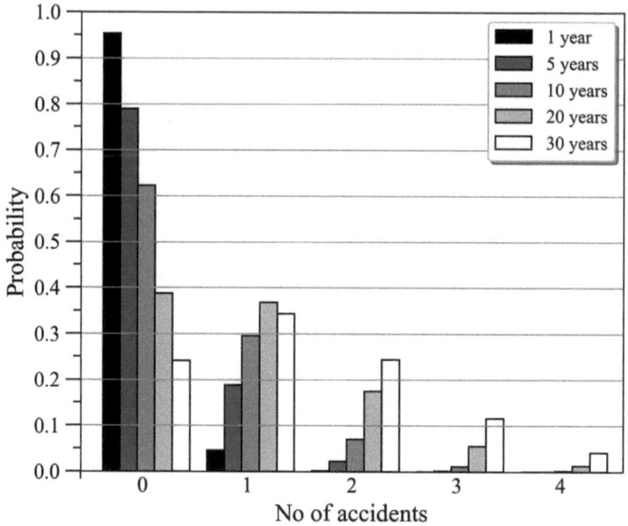

Fig. 8. Probability distributions of any fire accident for different time perspectives

The results suggest that the most probably scenario is no fire accident even in longer time horizons (0.95 for one year, 0.78 for five years and 0.62 for ten years). However, the probability of a single fire accident isn't not that low. Even for one year, it reaches about 0.05, and grows for longer perspectives. It is 0.19, 0.30, 0.34 and 0.36 for five, ten, twenty and thirty years, respectively. Having in mind that every tunnel fire in an unfortunate turn of events could pose a great threat to people trapped in the tunnel, such possibility must be taken seriously. This concerns especially the considered tunnel, which is a urban tunnel with a high probability of traffic jam, which may cause many people to get stuck inside. The probability that in the periods considered more than two fires break out quickly drops.

The expected numbers of fire accidents in different periods are shown in Table 1. They are np products expressed by formulas (3) and (4) and calculated for these periods.

Table 1. Expected numbers of fire accidents in different periods.

Expected number of fire accidents	Period		
	1 year	5 years	10 years
Non HGV	0.034	0.170	0.340
HGV	0.013	0.067	0.135
Total	0.047	0.237	0.475

Figure 9 shows the probabilities of the occurrence of a catastrophic tunnel fire in selected time perspectives. Such a fire must involve an HGV, and in addition, the fire power must be high enough (more than 20 MW). The latter condition diminishes significantly the probability due to the assumed exponential distribution of fire powers.

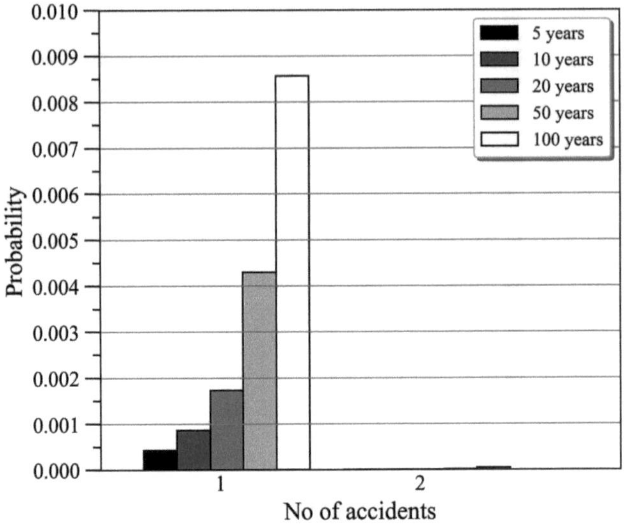

Fig. 9. Probability distributions of a catastrophic fire accident for different time perspectives

The probabilities, even for very long time perspectives, seem to be very low. But they refer to a catastrophic accident, with many potential casualties (Król and Król, 2021). In shorter periods (up to 20 years), the probability doesn't exceed a small fraction of 1%, but can't be neglected in light of possible severe consequences. Happily, the probability of two such catastrophic fires appears to be extremely low.

Although it is very unlikely that a fire accident will happen just today (probabilities for a single day are $8.7 \cdot 10^{-5}$, $2.3 \cdot 10^{-6}$ and $8.9 \cdot 10^{-5}$ for non trucks, trucks and in total, respectively), it seems to be interesting to determine the hourly probability distribution of such an event. It is important to find times of the day that are especially vulnerable to the fire outbreak. This stage of the work constitutes the essence of the research. Figure 10 presents the distribution of the relative probability of such an event hour by hour. The values are related to the maximum value of $1.1 \cdot 10^{-5}$, which corresponds to the probability of a fire accident on workdays between 3 pm and 4 pm. Since the traffic patterns for workdays and weekends have appeared to be different, three diagrams are prepared: for workdays, weekends, and the averaged one.

The curves of the relative probability roughly resemble the curves of the traffic intensity. This observation agrees with expectations, because the number of fire accidents depends on the number of travels through the tunnel. On workdays, there are two peaks visible, but the afternoon peak is slightly higher, and reaches its maximum on 3 pm. The curve for weekends, after a relatively calm morning, forms a plateau that ends after 9 pm. What is important is that the probability of a fire accident on weekends is the same than on workdays in the evening. Generally, in the period between 6 am and 8 pm on workdays, and between 10 am and 8 pm on weekends, the relative probability of a fire accident is significant. Nights are expected to be rather calm.

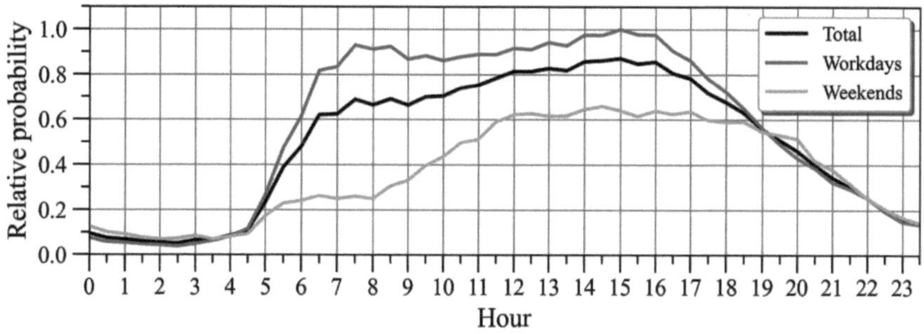

Fig. 10. Hourly distribution of the relative probability of a fire accident

However, it must be noted that it is just probability. The recent fire accident in the road tunnel 'Emilia' in Laliki on the expressway S7 happened about midnight on Feb 25th 2024. It was an HGV fire that completely destroyed the vehicle.

The share of HGV on weekends is negligible, so it makes sense to split the information on the hourly distribution of the probability of a tunnel fire between HGVs and other vehicles only for workdays. This is shown in Fig. 11. The values are related to the same reference value. The curve for workdays is presented again to make a comparison possible.

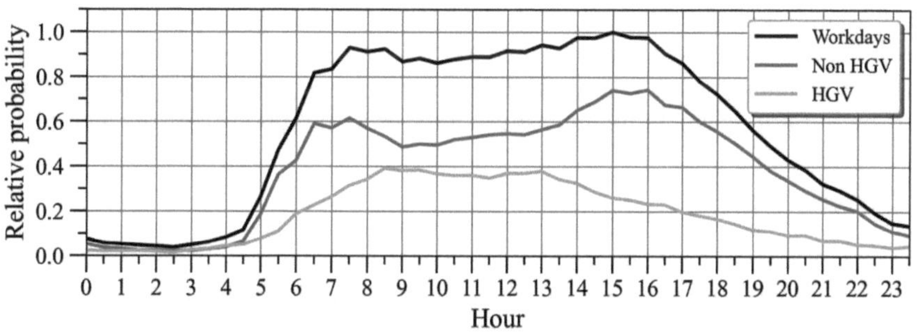

Fig. 11. Hourly distribution of the relative probability of a fire accident om workdays split for vehicle types

There are two peaks for the probability distribution for vehicles other than trucks, which mimic peaks corresponding to rush hours. Meanwhile, the curve of the probability for HGVs forms a plateau between 8:30 am and 1:00 pm. In this period, the value of the fire accident probability keeps an almost constant value of 0.4. This value is only slightly lower than the probability of other vehicles, despite the significantly lower share of HGV.

At the end of the work, the issue of the impact of the share of HGVs on the probability of a fire accident was examined. As mentioned earlier, tunnel fires are more often involving trucks than other vehicles. Figure 12 presents this dependence for different values of traffic intensity for a period of year.

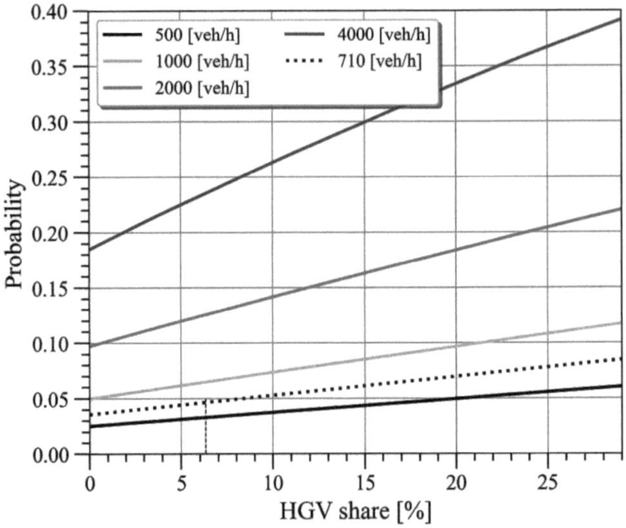

Fig. 12. Dependence of the probability of a tunnel fire on the share of HGVs

The diagram also shows this dependence for the examined tunnel (dotted line). The actual average share of HGV is marked by a thin dashed line. It is visible that the probability of a fire accident grows with the HGV share. This growth is much steeper for higher traffic intensities. It is important, because for two urban road tunnels in Poland (Warsaw and Katowice), traffic intensities of the order of 4000 veh/h are reported.

5 Conclusions

The research is a case study for a urban road tunnel under Martwa Wisła river in Gdańsk. The literature statistics on fire accidents in road tunnels combined with data on traffic intensity and its generic structure for the tunnel examined were used to assess the likelihood of a tunnel fire in different time horizons.

It has appeared that a single fire accident is probably in the ten years time horizon. The probability that a fire accident occurs on a single day is $8.9 \cdot 10^{-5}$, which may seem a low value. The analyzed reports suggest that trucks are much more prone to be involved in a fire accident than other vehicles, thus the total probability of such accident strongly depends on their share.

The hourly distribution of the fire risk was shown to be very uneven, because it depends on the traffic intensity and rapidly grows with the share of HGV. On workdays two peaks are visible, which correspond to rush hours. On weekends the probability is clearly lower because traffic is less intense in general and the HGV share is negligible.

The predictions were carried out basing on actual traffic characteristics, and one can expect that the traffic will increase in the future. Since the tunnel is located near the harbor area in Gdańsk, so the share of HGV could also increase soon. Therefore, the probability of a fire accident in the tunnel can be significantly higher. In this light, it is very important to keep the rules for the transport of hazardous materials.

Acknowledgments. The authors thank Road and Greenery Management (ZDiZ) Gdansk for the opportunity of the research and sharing the data.

References

1. Amundsen, F.H., Engebretsen, A.: Studies on Norwegian road tunnels II. An analysis on traffic accidents in road tunnels 2001–2006. Vegdirektoratet, Roads and Traffic Department, Traffic Safety Section, Raport no: TS4-2009. Statens Vegvesen, Oslo (2009)
2. Lu, J.J., Xing, Y., Wang, C., Cai, X.: Risk factors affecting the severity of traffic accidents at Shanghai river-crossing tunnel. Traffic Inj. Prev. **17**(2), 176–180 (2016)
3. Beard, A., Carvel, R.: The Handbook of Tunnel Fire Safety. Thomas Telford Ltd., London (2005)
4. Zhu, R., Zhang, D.M., Huang, Z.K., Guo, X.Y., Gan, B.L., Zhou, W.D.: Traffic-based resilience assessment on urban road tunnel affected by fire accident. 161, art no 106543 (2025)
5. Hua, N., Tessari, A., Khorasani, N.E.: Characterizing damage to a concrete liner during a tunnel fire. Tunn. Undergr. Space Technol. **109**, art no 103761 (2021)
6. Shen, Y., Ling, J., Wang, W., Zhu, H., Yan, Z.: 3D numerical investigation on response of shield tunnel under combined effects of fire and structural loading. Tunn. Undergr. Space Technol. **128**, art no 104659 (2022)
7. Shen, Y., et al.: Thermo-mechanical analysis of fire effects on the structural performance of shield tunnels. Tunn. Undergr. Space Technol. **132**, art no 104885 (2023)
8. Zhou, Z., Shen, Y., Yang, W., Yan, Z., Zhu, H.: Coupled CFD-FEM methodology for structural fire safety of composite lining shield tunnel. Tunn. Undergr. Space Technol. **150**, art no 105834 (2024)
9. Zhang, D.M., Zhu, R., Huang, Z.K., Yu, L.: Probabilistic analysis of fire casualties in an urban traffic tunnel considering the uncertainty of vehicle queue. Tunn. Undergr. Space Technol. **162**, art no 106619 (2025)
10. Król, A., Król, M.: The factors determining the number of the endangered people in a case of fire in a road tunnel. Fire Saf. J. **111**, art no 102942 (2020)
11. Król, A., Król, M.: Numerical investigation on fire accident and evacuation in a urban tunnel for different traffic conditions. Tunn. Undergr. Space Technol. **109**, art no 103751 (2021)
12. Feng, J., Sun, Y., Sun, Y.: Multiscale network based on feature fusion for fire disaster detection in complex scenes. Expert Syst. Appl. **240**, art no 122494 (2024)
13. Zhao, H., Jin, J., Liu, Y., Guo, Y., Shen, Y.: FSDF: a high-performance fire detection framework. Expert Syst. Appl. **238**, art no 121665 (2024)
14. Szewczyński, K., Król, A., Król, M.: Should we expect a disastrous fire accident in an urban road tunnel? literature data review and a case study for selected tunnels in Poland. Sustainability. **13**, art no 6172 (2021)
15. Król, A., Król, M., Węgrzyński, W.: A study on airflows induced by vehicle movement in road tunnels by the analysis of bulk data from tunnel sensors. Tunn. Undergr. Space Technol. **132**, art no 104888 (2023)

16. Rattei, G., Lentz, A., Kohl, B.: How frequent are fires in tunnels – analysis from Austrian tunnel incident statistics. In: 7th International Conference Tunnel Safety and Ventilation, Graz (2014)
17. Willman, C., Defert, R.: Statistical analyses of breakdowns, accidents and fires in road tunnels in France. In: 7th International Symposium on Tunnel Safety and Security, Montreal (2016)
18. PIARC, Design Fire Characteristics for Road Tunnels, Norman Rhodes, Paris (2017)
19. Aralt, T.T., Nilsen, A.R.: Automatic fire detection in road traffic tunnels. Tunn. Undergr. Space Technol. **24**, 75–83 (2009)

Evaluation of the Effectiveness of Traffic Calming Measures on the City's Street and Road Network

Olha Kholodova(✉), Maryna Buhaiova, Andrii Sakhno, and Oleh Kholodov

Faculty of Transport Systems, Kharkiv National Automobile and Highway University, Kharkiv, Ukraine
olgakholodova280781@gmail.com, kazmar2383@gmail.com, Mr.andrey2050@gmail.com, kholodovoleg1408@gmail.com

Abstract. The paper proves the relevance of improving traffic management in connection with the growth of motorization and the insufficient development of the street and road network of cities. Traffic calming devices are considered as a tool for improving the safety and comfort of the urban environment. An analysis of the effectiveness of the introduction of various types of traffic calming devices at intersections and pedestrian crossings is carried out. It is proven that the combination of traffic lights and traffic calming improves safety, although it may cause additional delays. It is established that the most effective means at intersections are raised roads. The obtained relationships between traffic volume and performance indicators can be used for a preliminary assessment of the impact of traffic calming measures. The results provide a basis for further research on the integrated application of traffic calming measures in the urban environment.

Keywords: traffic calming devices · road hills · elevated pedestrian crossings · elevated intersections · Berlin cushions · transport modelling

1 Introduction

The annual growth in the number of cars and traffic volumes leads to an increase in traffic intensity, which, in the context of existing urban development, leads to overloading of the street and road network (SRN), increased delays, an increase in the number of road traffic accidents (RTAs) and a deterioration in the environmental situation in Ukrainian cities.

These problems are particularly acute at intersections, which are the main points of concentration of RTAs. According to the National Highway Traffic Safety Administration, a significant proportion of accidents occur at these junctions. In Ukraine, about 40% of RTAs occur at intersections, which confirms their critical role in ensuring road safety.

The historically developed SRN of most Ukrainian cities, with narrow streets and uneven road layout, does not meet modern transport needs. Increasing traffic intensity causes overloading of the SRN, reduces the speed of traffic flows (TF), and increases

E. Macioszek et al. (Eds.): TSTP 2025, LNNS 1769, pp. 55–74, 2026.
https://doi.org/10.1007/978-3-032-14094-4_5

noise, air pollution, and fuel consumption. To improve traffic efficiency and safety, it is necessary to modernize transport infrastructure using modern technologies.

In Ukraine, considerable attention is paid to the long-term development of the SRN, taking into account the growth of the urban population and changes in the transport situation. The introduction of modern architectural, planning, and organizational measures helps to improve the comfort and safety of road users. An important area of improving the road situation is the use of traffic calming devices (TCDs), which have proven to be an effective means of improving road safety [1].

Traffic calming is a set of engineering, educational, and legal measures aimed at reducing vehicle speeds and improving the safety of road users. The main goal is to reduce the number and severity of RTAs and create a comfortable public space. In global practice, traffic calming is a key element of transport planning and is implemented through the 'three E's' approach (engineering, education, enforcement), contributing to the harmonious development of urban infrastructure. As a result of properly applied traffic calming measures, the average vehicle speed is reduced to a safer level, the number of RTAs and the severity of their consequences are reduced, conditions for pedestrians, cyclists, and public transport are improved, and transit traffic is reduced. This type of measure was rarely considered in the national urban planning literature until recent decades, but is now gradually gaining popularity in the field of transport planning.

That is why the purpose of this study is to provide certain recommendations for improving road safety at the intersection and pedestrian crossing by selecting the most effective TCDs to be implemented at them by analyzing the performance indicators of their application.

Section 2, 'Literature Review,' contains a detailed review of the literature on various TCDs, their classification, advantages, and disadvantages, with a special focus on Berlin speed cushions (BSCs). Section 3, 'Research Methodology,' describes the research methodology, including the selection of individual cases on the SRN of the city of Kharkiv and the use of simulation modelling in PTV VISSIM to evaluate various traffic calming measures. Section 4, 'Results and Discussion,' discusses the results of the modelling, compares traffic performance indicators in several scenarios, and highlights the advantages of BSC. Section 5, 'Conclusions,' summarizes the research results, generalizes the main conclusions, and highlights the potential of BSC as a balanced solution for urban traffic safety and efficiency.

2 Literature Review

Traffic calming has long been used worldwide to control speed and traffic intensity, but in Ukraine, until 2020, the "sleeping policeman" was mainly used. Traffic calming aims to increase the safety and comfort of pedestrians and cyclists, especially in urban areas and narrow streets with high pedestrian activity.

An analysis of literary sources has shown that the number of traffic calming is very large, and each means has its advantages and disadvantages [2]:

1. Road hills (RHs) reduce vehicle speed, which helps reduce the risk of accidents and increases drivers' reaction time to dangerous situations on the road. At the same time, RH increases vehicle travel time and fuel consumption, which leads to an increase in

negative environmental impact and worsens the condition of the road surface. Their inappropriate placement can contribute to an increase in traffic congestion on the SRN.

2. Elevated pedestrian crossings (EPCs), along with ensuring pedestrian safety and reducing the risk of injury when crossing the road, reduce the speed of vehicles, which causes congestion at the approaches to pedestrian crossings. However, reducing speed helps reduce the risk of accidents. The presence of an increase can changes the usual movement of drivers, lead to instability in their movement, and require additional attention on their part. EPC also requires improved access for people with disabilities.
3. Elevated intersections (EIs), like EPC, reduce the speed of vehicles at the intersection, thereby improving traffic safety for pedestrians and cyclists. However, by forcing drivers to drive slower, elevated intersections increase travel time on the SRN, which is ineffective in solving traffic management problems regarding network management. The high cost of constructing elevated intersections can be a rather expensive measure for our country, particularly due to the need for additional space and materials. It should also be noted that the comfort of movement for public transport passengers deteriorates due to increased repulsive force and pushing. Elevated intersections require constant maintenance, i.e., maintenance costs, which may be associated, for example, with snow and ice removal in winter.
4. Chicanes have become most popular in places with many pedestrians, especially children. They also reduce vehicle speeds, which helps reduce the number of accidents and the severity of their consequences. Chicanes reduce noise levels and environmental pollution and help reduce road maintenance costs. However, they make it difficult for some types of vehicles, namely large trucks and buses, to increase vehicle travel times, especially during peak hours, and inconvenience cyclists and motorcyclists.
5. Mini-rings, as proven by most scientific studies, are one of the most effective ways to increase the throughput capacity of an intersection [3], especially during rush hour, when other intersections can cause traffic jams. Like the previous TCDs, mini-rings reduce vehicle speed and, therefore, the risk of accidents. In addition, they reduce waiting times at traffic lights, as traffic flows almost without stopping, and generally improve the aesthetic appearance of the intersection, as the island can be decorated with greenery or sculptural structures. However, mini-rings may require road widening and additional lighting. They complicate the movement of cyclists, pedestrians, and large vehicles, which can cause traffic jams. In forest belts, they help reduce the incidence of animal collisions.
6. Channeling TF allows reducing the number of conflict points on the elements of the SRN, since the movement occurs in specific directions. For each direction, it is possible to plan the rational operation of traffic light objects, which reduces the waiting time of vehicles at the traffic light. When channeling, the visibility of traffic light objects, road signs, markings, etc., can be increased; however, the consequence is an increase in the costs of their arrangement and subsequent maintenance. The disadvantage of channeled traffic is the need for additional spatial resources to organize the separation of flows and reduce the capacity of the road, which is also associated with the separation of flows.

7. Intersection blocking reduces its capacity, creating additional congestion. The time drivers spend crossing the intersection depends on the configuration of the obstacles and barriers, making it difficult for pedestrians and cyclists to pass through the intersection. However, such blocking reduces the likelihood of accidents, namely collisions, and reduces the number of accidents with pedestrians and cyclists. Control over vehicle traffic at the intersection also improves.
8. Chokers reduce traffic speed, reducing the risk of accidents and providing excellent safety for pedestrians and cyclists. The risk of vehicles crossing road markings is also reduced. The disadvantages include that chokers increase the risk of congestion during peak hours with large volumes of vehicle traffic, which reduces the capacity of the SRN elements and makes it impossible to move large vehicles.
9. Road centerline inserts improve traffic safety by improving visibility under challenging conditions, creating additional space for traffic, and reducing the likelihood of collisions. They contribute to a smooth flow of vehicles and reduce congestion by allowing slow vehicles to overtake. However, their installation is expensive, especially in large areas, and high traffic volumes can cause noise and vibration. Inserts require regular maintenance and can impair visibility on sharp turns or in areas with limited visibility.

Among the modern TCDs, the so-called Berlin cushions are particularly noteworthy - engineering structures that combine efficiency in reducing the speed of passenger cars with maintaining the smooth flow of public transport. First introduced in Berlin in the late 1980s, these elements are an example of an adaptive approach to urban traffic management that prioritizes the safety of vulnerable road users without significantly disrupting transport logistics. Due to their design features - a smaller width compared to bus lanes - Berlin cushions provide an effective speed reduction of 25–40 km/h for passenger vehicles, while allowing large vehicles to cross the section without compromising comfort. This balance between safety and capacity makes Berlin cushions a promising solution for implementation in Ukrainian cities, particularly on high-density streets, near educational institutions, or within public transport routes. At the same time, the current regulatory documents of Ukraine do not currently provide for such structures, which complicates their widespread use in practice, but the available international experience shows the feasibility of their regulatory adaptation in the context of the national SRN [4].

The analysis of TCDs allows us to conclude that they can effectively control the traffic speed on city streets. Still, they should be used with understanding, carefully justifying a particular type of device.

Traffic calming is achieved by changes to the SRN and technical measures, in particular, eliminating transit traffic through the city center by converting through streets into loops or dead ends. Speed limits are also introduced, which reduces conflicts between traffic and pedestrians. Street landscaping and design play an essential role in influencing traffic patterns. Zones are often served by public transport, which is given priority, as in Strasbourg and Dijon (France). The organization of street space encourages a decrease in vehicle speed, even at the expense of the capacity of the SRN [5].

The range of technical techniques and TCDs is extremely wide: organization of parking areas; channelization, lane separation; use of sharp turns; deviation of the trajectory at the intersection; use of road hills ("speed bumps"); narrowing of streets; restriction of access; use of roadway elevations with a change in texture; application of restrictions on vehicle sizes; use of dead-end streets; redevelopment of streets; use of roundabouts; use of traffic control elements [2].

When determining the effectiveness of using traffic calming, the following main results are distinguished, which are achieved by this method: reduction of the number and severity of road accidents in cities at intersections and intersections; reduction of vehicle speed; reduction of transit traffic of vehicles in the city.

The concrete results of a successful traffic reduction policy can be illustrated by the example of the city of Bruges (Belgium), where a traffic calming was introduced in the historic city center back in 1992: a 30% reduction in traffic in the center; an increase in the speed of bus routes from 19 to 22 km/h; a 10% reduction in the number of vehicles entering the center (by 600 vehicles/h); a 33% increase in the number of residents using buses; a 20% increase in the number of residents using bicycles; a 36% reduction in the number of road accidents in the center. Data on the implementation of traffic calming in the city of Graz (Austria) also indicate positive results: in particular, speed limits significantly impacted traffic safety and contributed to a positive assessment of traffic calming among residents and drivers. The first major experiment in implementing a traffic calming zone in the United States was the Stevens Neighborhood in Seattle. The project was completed in early 1973. As a result, traffic intensity was reduced by 56%, and the number of accidents was reduced to 0.

TCDs are used as a means of redistributing traffic on the SRN. It should be noted that the use of calming zones and speed limit zones implies that the maintenance of vehicles is being carried out by other sections and elements of the SRN. This is well confirmed by the data of surveys carried out in the places where these measures are implemented. According to the sources, namely, the results of the analysis of data from 43 international studies showed that the use of traffic calming measures in cities reduces road accidents by 8–100%. At the same time, there was not a single case of an increase in the number of road accidents after the introduction of these devices, i.e., in all cases, positive results and feedback were obtained.

According to [2], there are 3 groups of road traffic control measures: 1) the construction of obstacles on the carriageway (implemented with the use of RHs, EPCs and EIs; 2) changing the traffic trajectory (implemented with the use of chicanes, mini-rings, channelization and blocking the traffic of the vehicle); 3) changing the width of the carriageway (implemented with the use of chokers and inserts along the road axis).

Given their design specifics, it is advisable to include Berlin cushions in the third group of traffic calming measures - those that involve changing the width of the carriageway. Despite the fact that this element is not currently mentioned in the current regulatory documents of Ukraine, a number of cities are already attempting to implement it in practice. For example, in the summer of 2023, Vinnytsia, Dnipro, and Odesa announced the abandonment of obsolete speed bumps in favour of more modern Berlin pillows. In Kyiv, similar decisions are still being discussed, due to both the lack of a regulatory framework and caution about integrating new engineering solutions into the existing transport infrastructure. However, given the positive experience of these cities, we can expect to see the practice of implementing Berlin cushions in other Ukrainian cities, particularly in large agglomerations with heavy traffic and the presence of school or pedestrian active zones. This could be an effective step towards adapting European approaches to safe urbanism to the Ukrainian street and road network, especially given the needs of post-war urban reconstruction.

An analysis of the current scientific literature demonstrates the growing interest in the effectiveness of traffic calming measures in the urban environment, in particular BSCs, as an engineering solution that combines speed reduction with minimal impact on emergency services. In a previous study, the authors analyzed the impact of BSCs on the operational speed of vehicles within the 30 km/h zone in Bari, recording up to 30% reduction in speed (V85) depending on the geometric characteristics of the cushions, and also identified changes in the trajectory of two-wheeled vehicles that require additional attention [6]. In a subsequent, more detailed study, the same authors critically assessed the limitations of previous work, focusing on the effect of cushion length (2.20 m, 2.70 m, 3.20 m) and contextual factors (time of day, day of week, traffic intensity). The results of the analysis of variance (ANOVA) confirmed a significant decrease in speed before and after the cushion (up to 12–13 km/h) and revealed greater efficiency in the morning. At the same time, longer cushions showed a slightly lower impact on driver behavior, which indicates the need for a comprehensive assessment of such measures, taking into account the specifics of the urban context [7].

As a result of research in [8] on the effectiveness of speed cushions as a measure of forced speed reduction in residential areas, it was found that they are somewhat less effective than traditional speed bumps in reducing speed, but have advantages in comfort for passengers of large vehicles (buses, special equipment), who can drive around or over them without significant vertical oscillation. The study covers the impact of cushion parameters (width, length, slopes), their placement, and the distance between them on speed, driver behavior, traffic flow, noise, vibration, passenger comfort, and occupant perception. It was concluded that 1600 mm wide pavements may not be effective enough for areas with a speed limit of 20 mph without additional measures.

The authors of [9] consider speed cushions to be one of the most effective means of physical TCDs, which are a modified form of road humps with channels for emergency services. According to a study conducted by the King County Department of Transportation (Washington State), the introduction of such cushions reduced vehicle speeds without significantly affecting the response time of fire and emergency vehicles. As part of the pilot project, ten speed bumps were installed on several streets in a residential area, and the results of the effectiveness assessment confirmed their feasibility as a compromise solution between road safety and the efficiency of rescue services. The authors also emphasize the need for further standardization of the design, color, and placement of such elements.

As part of the study of speed management on urban roads [10], we analyzed newsletters published by the Ministry of Transport of Quebec, which are an important source of empirical and analytical data on the implementation of TCDs. These materials consider various engineering measures aimed at reducing vehicle speeds and improving safety, in particular for vulnerable road users. The general fact sheet presents the methodology for the preliminary analysis that precedes the implementation of the measures, while the specialized fact sheets detail specific types of infrastructure solutions such as speed bumps, safety islands, EPCs, etc. Each bulletin contains an analysis of the effectiveness of the measure, its geometry, maintenance conditions (including winter), and potential safety impact. The authors note that the documents presented by the ministry are relevant for municipalities looking for evidence-based guidelines for implementing speed reduction policies on their territory.

In the context of creating a safe environment for pedestrians, it is becoming increasingly important to rethink the spatial distribution of streets, which involves reducing or transforming the space intended for motor traffic. In the scientific literature, in particular, in a study on integrated traffic management measures [11], two main approaches to vehicle speed limits are described: 'hard' - with the use of physical barriers, and "soft" - with psychological influence on drivers through visual or spatial signals. The authors analyze the use of vertical and horizontal elements (e.g., raised platforms, roadway narrowing, safety islands) integrated with landscaping or other architectural elements to reduce traffic speeds, improve pedestrian crossing conditions, and reduce the dominance of cars in urban space. It is proposed that such measures be evaluated using a multi-criteria approach, including not only effectiveness in reducing traffic speeds or volumes, but also pedestrian experience, maintaining access for residents and emergency services, impacts on drainage and winter maintenance, loss of parking spaces, and achievement of environmental goals, including noise and air pollution reduction.

A fairly large number of studies in the field of TCDs are aimed at comparing the number of road accidents before and after the introduction of various means. Thus, in [12], an analysis of data on fatal and injured road accidents on road sections showed that the introduction of vertical traffic deceleration reduces the number of fatalities and injuries by 60%; the placement of safety islands with horizontal markings and flexible reflective poles - by 72.7%; and the placement of safety islands on the main road - by 35.7%.

Interesting developments are presented in [13], where the best measures for slowing down traffic are a raised pedestrian crossing and narrowing of traffic lanes. At the same time, the authors insist that even better results can be obtained when more than one traffic calming measure is used along the street, with a distance between them not too large.

In [14], the same authors compare the speed reduction from the introduction of EPCs and the narrowing of traffic lanes. However, it is necessary to consider the driver's memory effect, which is maintained along the entire length of the narrowed roadway (for example, due to parking areas). In contrast, the driver does not maintain a slow speed after passing the EPCs.

Studies confirm the effectiveness of traffic calming for safety and ecology. They are considered a tool for improving the traffic situation [15], a combination of strategies for minimizing the impact of vehicles on the environment and society [16], as well as physical measures for correcting driving style and protecting pedestrians and cyclists [17]. The analysis showed that TCs are divided into 13 categories and can be combined with existing transport systems.

In [18], the effect of bicycle lanes on traffic speed was investigated. An analysis of over 9,000 vehicles showed that delineated lanes reduce maximum speed by 28% and right-turn speed by 21%. Straight-through traffic is slowed by 8%. The study confirms that cycle lanes improve cyclist safety and reduce the risk of accidents for all road users.

Paper [19] investigated the impact of traffic restrictions on communities in Singapore, in particular the Silver Zone programme, which improves the safety of the elderly. The analysis of the residents' responses revealed three groups of sentiments: 'Safety Enthusiasts', 'Uncomfortable Users', and 'Implementation Critics'. Despite their different views, they all generally supported the programme, recognizing its positive impact on safety and quality of life. The study emphasizes the importance of public involvement in improving the effectiveness of traffic calming.

Studies of traffic calming have considered traffic restrictions as a way to improve safety. For example, [20] has shown that lowering speed limits in cities (to 30 or 15 km/h) can reduce accidents while encouraging the use of public transport, as observed in Berlin.

In [21], the authors examine the impact of traffic calming on the livability of urban residential streets. Using georeferenced data on installing 1,187 traffic calming devices in Portland, Oregon, they test whether interventions affect local housing prices over the following years. It is shown that urban residents pay significant premiums to limit their exposure to cars, but the results on the overall impact on the price of TCDs are mixed. Only the most effective traffic-reducing measures have a significant effect on housing prices. The estimated elasticity of traffic flow is -0.07: projects that reduce traffic by 16% increase the value of houses on treated streets by 1%.

The aim of [22] is to assess the importance of urban traffic calming and to gain insight into the safety impact of local and regional traffic calming interventions, allowing for better understanding and, therefore, "better-informed" decisions about different schemes. The effect of traffic calming initiatives on safety was analyzed at the city level for large Hungarian cities, and an in-depth study was conducted for a more detailed case study of the Hungarian capital. The results show that traffic reduction initiatives are vital in improving city road safety. At the same time, the Budapest study explains the observed effects in detail, which allows for the development of more precise assessment tools.

Scientists compare the effectiveness of traffic calming. In [23], Spherical Cap Studs are recognized as an effective alternative to speed bumps for reducing the speed of two-wheeled vehicles with less discomfort for drivers. Their use in cities is expected. Research [24] confirms that TCDs are more effective in complex schemes. Four combinations of measures were analyzed in risky areas, which reduced speeds by 30–50%. The public positively assessed their impact, especially on rough surfaces. Policymakers are advised to combine measures to increase effectiveness.

The authors [25] emphasize the need for a systematic approach to selecting traffic calming for their effectiveness. They note that various measures (speed bumps, road signs, widening of shoulders, etc.) have been implemented in several Asian cities, but many have failed due to a lack of proper planning. In Japan, the implementation of such measures is often accompanied by non-compliance with standards, which reduces their effectiveness and popularity among the population. Research shows that only well-designed health programs gain community support and bring significant benefits.

There are also studies on the impact of traffic calming on cities' environmental load. For example, in [26], traffic calming can negatively affect vehicle fuel consumption and emissions if drivers use aggressive acceleration levels to speed up their journey. Therefore, by eliminating sharp acceleration maneuvers through driver training, significant savings in vehicle fuel consumption and emissions can be achieved.

Research [27] showed that speed bumps effectively reduce vehicle speed, and their effect remains stable over time. The optimal height is 5–7 cm in dangerous areas and 3 cm in less risky areas. Research [28] confirmed the effectiveness of speed bumps of different sizes, and research [29] found that their effect extends to $\pm$ 30 m. Despite their prevalence on local roads, they have a more pronounced slowing impact on highways, which makes them a reasonable means of improving pedestrian safety.

Domestic research of traffic calming is primarily advisory in nature and is based on foreign experience. In particular, EPCs are considered adequate for pedestrian safety, although their installation is costly [30]. Research [31] notes the growth of traffic calming use in Ukrainian cities, particularly for speed control and transit traffic restrictions. Much attention is paid to mini-rings' effectiveness, but the need to study their design and methods for assessing capacity is emphasized. The use of transport modeling is recommended for analyzing the safety and effectiveness of mini-rings at the design stage [32].

That is why, within the framework of this work, a study will be carried out on the application of TCDs of the 1st group on various elements of the SRN using transport modelling, which has recently gained considerable popularity in solving traffic management problems. If micro-level analysis is required, in particular, when planning traffic at a single intersection, road section, group of intersections, or when visualizing changes for the public in the event of reconstruction or construction of new infrastructure facilities, PTV Vision VISSIM software is used. In addition, the study provides for a comparative analysis of the effectiveness of Berlin cushions, which belong to the Group 3 means, in order to comprehensively assess their impact on traffic parameters and the feasibility of using them in the urban environment.

3 Research Methodology

To evaluate the effectiveness of traffic management tools, this study analyses the implementation of different traffic management methods at one of the most dangerous intersections and one of the unregulated pedestrian crossings in Kharkiv (the second most populated city in Ukraine with an extensive SRN, combining dense residential development, heavy traffic and a high concentration of vulnerable road users). That is why the results obtained on the example of Kharkiv are representative of other large Ukrainian cities. The study is aimed at developing universal approaches to planning road safety measures adapted to the national context, with the prospect of scaling them up as part of the state transport policy.

The intersection of A. Oshchepkova Street (a city street) and Brigady Khartiya Street (a main street of district significance) are a straight type with organized two-way traffic and active pedestrian traffic (Fig. 1). According to road accident statistics, the main cause of accidents at this intersection is limited visibility. Despite the introduction of traffic lights in 2020, based on numerous public complaints, the accident rate remained high, in particular due to the location of small architectural forms that limit visibility and impede traffic, especially during martial law power outages. Cargo operations near the intersection further complicate the situation. Improving safety requires improving visibility, which is only possible by dismantling some of the small architectural forms, but this measure has socio-economic and legal constraints, including the availability of permits for occupied land plots.

Therefore, to improve road safety at this junction, taking into account the existing road conditions, it is necessary to reduce the speed of the vehicle. With a decrease in speed, the normative value of visibility [33] will also decrease, which will correspond to the existing state of the intersection.

Reducing the speed can be implemented with the use of: appropriate road signs (ineffective measure); use of traffic lights (ineffective measure); use of traffic control equipment. Each of these measures has its advantages and disadvantages. We will try to determine which of these options is more effective using simulation modelling in PTV Vision VISSIM. In the course of experimental studies, the intensity of vehicle traffic will change. It will vary from the base intensity to + or –20% in increments of 10%.

So, let's consider the following options for modelling the TF at the intersection: a controlled intersection (CI); RHs on the carriageway; EPCs on the carriageway; EIs on the carriageway; BSCs.

After modelling various traffic management options, the criteria for traffic management were analyzed. The most influential criteria for evaluating the developed measures were selected, and the level of service (LOS) was determined: average vehicle delay, number of vehicle stops, average length of traffic jam, and maximum length of traffic jam.

The objective function of the RTM options evaluation criterion at the intersection has the form

$$\begin{matrix} CI \\ RHs \\ EPCs \\ EIs \\ BSCs \end{matrix} = \left\{ \begin{matrix} N_{STOP}^{vehicle} \\ t_{average\ delay} \\ L_{traffic\ jam} \end{matrix} \right\} = f(N) \rightarrow \min, \quad (1)$$

where *CI* - controlled intersection; *RHs* - road hills; *EPCs* - elevated pedestrian crossings; *EIs* - elevated intersections; *BSCs* - Berlin Speed Cushions; *N*- TF intensity, vehicles/hour; $N_{STOP}^{vehicle}$ - number of vehicle stops, units; $t_{average\ delay}$- average vehicle delay, s; $L_{traffic\ jam}$- traffic jam length, m.

To determine the type and parameters of the models, a modern computer program for processing statistics, Statgraphics Centurion, was used, which allows for describing the obtained statistics with different kinds of mathematical functions. The results of experimental data processing are given in Table 1.

Let us consider the case of using TC at a pedestrian crossing. For the study, we will choose typical pedestrian crossings near secondary educational institutions in Kharkiv, where we see typical conditions: a residential street (Vasylya Melnykova St.) and a main street of district significance (Oshchepkova St., Biblika St., Hromadyanska St.) have 1 lane in each direction of movement, each 3.0 to 4.0 m wide. The pedestrian crossing is located opposite the main entrance to the educational institution. The visibility of

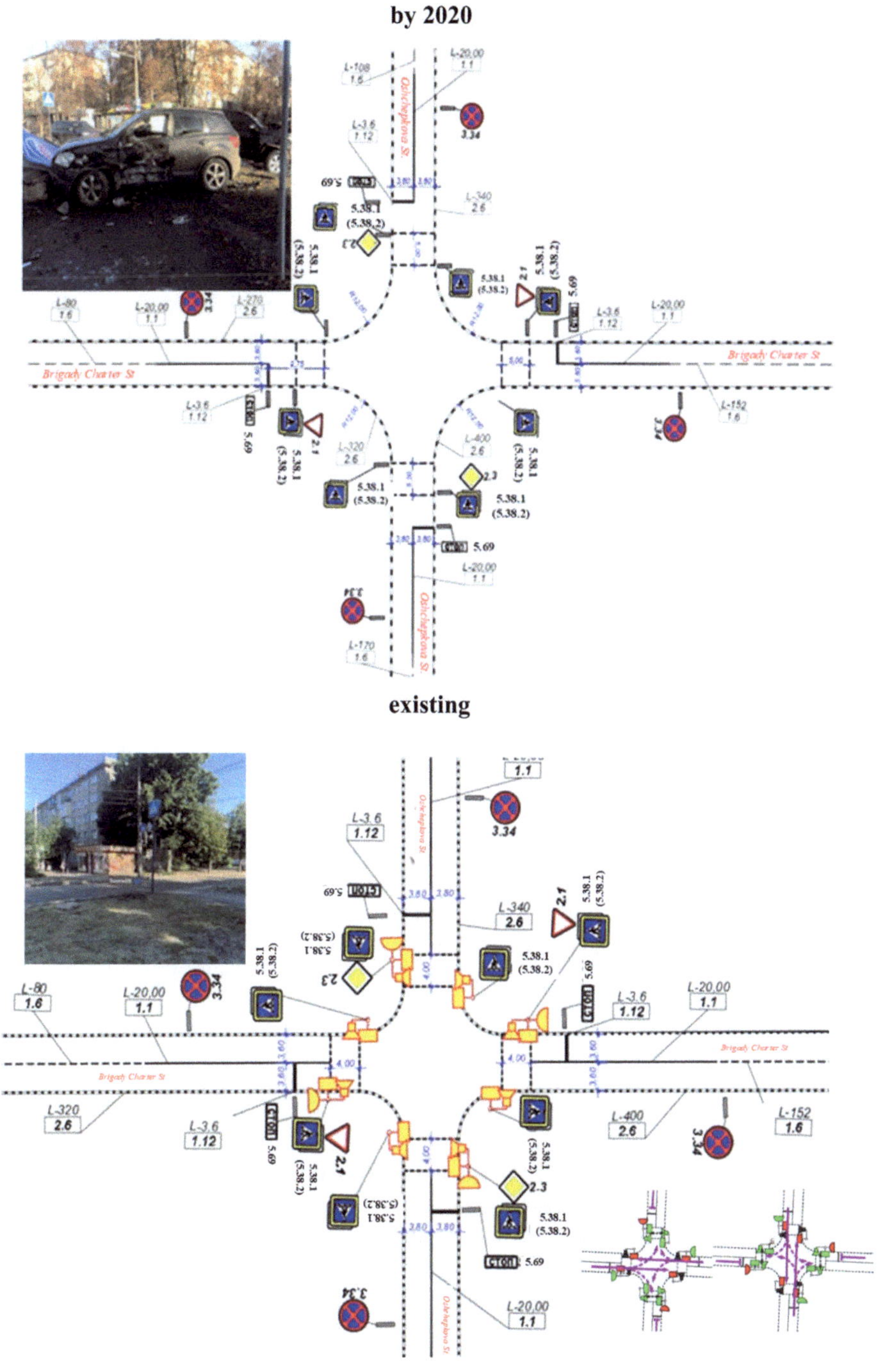

Fig. 1. Schemes of technical means of regulation and phase-by-phase departure at the intersection of A. Oshchepkova St. - Brigady Charter St., Kharkiv

road signs 5.38.1 and 5.38.2 - Pedestrian crossing is limited due to tree planting. There is no road marking indicating a pedestrian crossing. The street's straight section provokes vehicles' movement at increased speeds. On Oshchepkova St., Biblika (before the war) organized public transport routes: trolleybus and bus. In the morning and evening, when children are brought and picked up, many vehicles are parked near educational institutions, which additionally causes restrictions on the visibility of the pedestrian crossing.

To improve road safety, by DSTU 4123:2020 "Road Traffic Safety. Traffic Calming Devices. General Technical Requirements" [34], traffic calming facilities are arranged near educational institutions - RH or EPC. To pass these artificial road irregularities, drivers must reduce the vehicle's speed regardless of whether pedestrians are on or in the pedestrian crossing zone. Here, the question arises of studying the impact of the use of traffic calming facilities on the functioning of the transport network.

As criteria for assessing the effectiveness of the use of traffic calming facilities, we will use the following indicators: average delay time per vehicle, s ($\overline{T_{zup}} \to$ min),average length of the traffic jam, m ($\overline{L_{zat}} \to$ min), average number of vehicle stops, units ($\overline{N_{zup}} \to$ min). To assess the effectiveness of the use of traffic calming facilities at a pedestrian crossing, the model analysis settings were made in the PTV Vision VISSIM software.

As a prototype, a pedestrian crossing was taken to study traffic and pedestrians.in front of the Technological Lyceum №9 on Hromadyanska St. A pedestrian crossing model was developed for three options: option 1 - organization of a pedestrian crossing using road signs and road markings; option 2 - organization of a pedestrian crossing using an EPC, road signs, and road markings; option 3 - organization of a pedestrian crossing using an RH road signs and road markings; 4 - application of BSCs in combination with a pedestrian crossing.

The calibration of the VISSIM simulation model was evaluated using the Geoffrey E. Havers (GEH) statistic, which quantifies the difference between simulated and observed traffic volumes. According to standard practice, GEH values below 5 for at least 85% of volume checkpoints are generally considered acceptable for model calibration. The GEH statistic, a measure comparable to the chi-squared test [35], is calculated as follows:

$$\mathrm{GEH} = \sqrt{\frac{2(M - C)^2}{M + C}}, \tag{2}$$

where M represents the simulated traffic volume and C denotes the observed traffic count from field data.

In this study, all calculated GEH values were found to be below the threshold of 5, confirming that the simulation model meets the established calibration criteria.

When analyzing the methods of traffic calming and applying TCDs for this, it was determined that the parameters for assessing the functioning of SRN, namely, pedestrian crossings. Are affected by such input characteristics as the incoming TF intensity and the incoming pedestrian flow intensity. It is also known that the values of these intensities change significantly during the hours of the day. Therefore, to vary the model's input parameters within this work's framework. The following indicators were determined: TF intensity and pedestrian flow intensity.

Table 1. Results of experimental data processing in the Statgraphics program

Traffic management options at intersections	Dependency	Correlation coefficient	Durbin-Watson statistic	Mean absolute error
CI	$N_{STOP}^{vehicle} = 5,1 - 0,01 \cdot N^{*} + 0,00006 \cdot N^{2}$	0.88	3.57 (P = 0.93)	0.064
	$t_{average\ delay} = 70,8 - 0,132 \cdot N + 0.000069 \cdot N^{2}$	0.95	2.816 (P = 0.2095)	0.75
	$L_{traffic\ jam} = 21,9 - 0,045 \cdot N + 0,000025 \cdot N^{2}$	0.98	3.174 (P = 0.565)	0.159
RH	$N_{STOP}^{vehicle} = 0,13 \cdot e^{0,202 \cdot N}$	0.98	2.37091 (P = 0.38)	0.06
	$t_{average\ delay} = 49,3 - 0,105 \cdot N + 0.000064 \cdot N^{2}$	0.94	2.76416 (P = 0.17)	0.75
	$L_{traffic\ jam} = 53,3 - 0,12 \cdot N + 0,00007 \cdot N^{2}$	0.94	2.8274 (P = 0.22)	0.57
EPC	$N_{STOP}^{vehicle} = 0,18 \cdot e^{0,002 \cdot N}$	0.97	2.55008 (P = 0.49)	0.05
	$t_{average\ delay} = 1,2 \cdot e^{0,002 \cdot N}$	0.97	2.68434 (P = 0.57)	0.06
	$L_{traffic\ jam} = 53,3 - 0,12 \cdot N + 0,00007 \cdot N^{2}$	0.94	2.9 (P = 0.23)	0.58
EI	$N_{STOP}^{vehicle} = 6,1 - 0,013 \cdot N + 0,00008 \cdot N^{2}$	0.99	3.16238 (P = 0.55)	0.02
	$t_{average\ delay} = 77,89 - 0,17 \cdot N + 0,0001 \cdot N^{2}$	0.99	3.55963 (P = 0.93)	0.77

(continued)

Table 1. (*continued*)

Traffic management options at intersections	Dependency	Correlation coefficient	Durbin-Watson statistic	Mean absolute error
	$L_{traffic\,jam} = 75,77 - 0,179 \cdot N + 0,00011 \cdot N^2$	0.99	2.65844 (P = 0.12)	0.38
BSC	$N_{STOP}^{vehicle} = 13,2 - 0,029 \cdot N + 0,000017 \cdot N^2$	0.9	2.67592 (0.1242)	0.24
	$t_{average\,delay} = 94,266 - 0,21 \cdot N + 0,000125 \cdot N^2$	0.98	2.6154 (0.0968)	1.1132
	$L_{traffic\,jam} = 61,24 - 0,142 \cdot N + 0,000086 \cdot N^2$	0.96	2.78464 (0.1877)	1.17

Note: $^*N$ - TF intensity.

It is determined that the intensity of the TF (N_t) will vary in the range from 150 cars/h to 750 cars/h with a step of 150 cars/h. and the intensity of the pedestrian flow (N_p). From 150 pedestrians/h to 550 pedestrians/h with a step of 100 pedestrians/h. By enumerating possible variations of intensities. we find the number of modeling options: 5 options for changing the TF and 5 options for changing the pedestrian flow. Thus. it is necessary to perform modeling of 25 options for input parameters of intensities for each of the three modeling scenarios. This choice of ranges and steps is justified by the need to cover typical traffic intensity values characteristic of similar urban crossings throughout the day, including both low and high traffic periods. The selected step allows obtaining a sufficient number of points for a detailed analysis of the dependence of the SRN performance indicators on input intensities, while ensuring a reasonable amount of modelling within the scope of this study.

Figures 2 and 3 demonstrate the results of internal validation of models for selected scenarios with the implementation of traffic calming measures in the absence of field data. The consistency of the results obtained with theoretical expectations confirms the adequacy of the constructed model. The graphs reflect the logical increase in negative traffic indicators (delays, queue lengths, number of stops) depending on the intensity, which corresponds to the predicted effects of the implementation of the relevant measures.

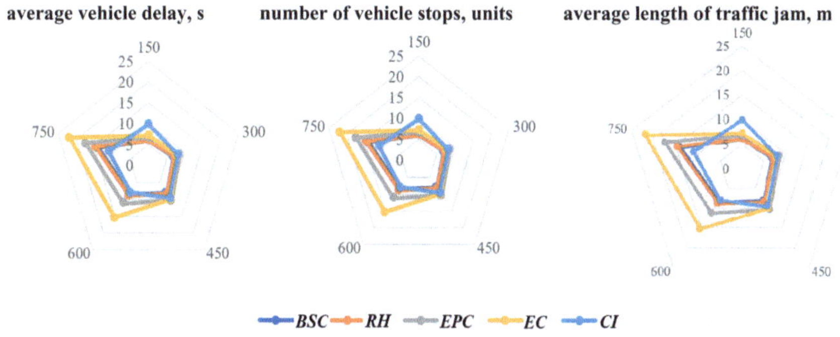

Fig. 2. Comparison of the proposed options by evaluation indicators

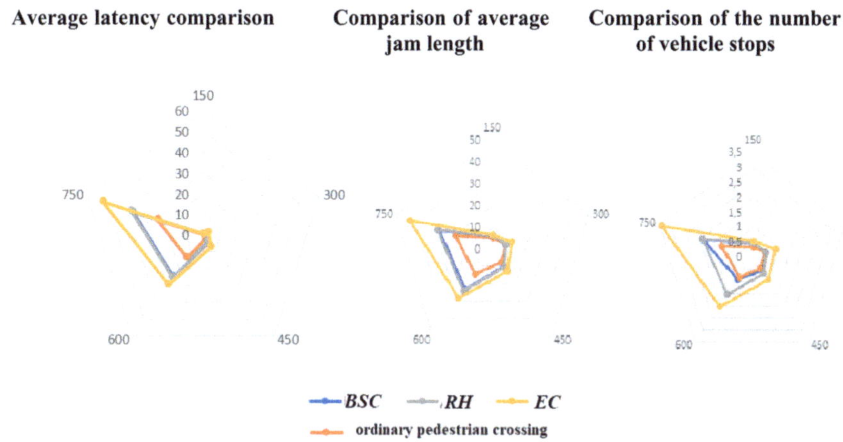

Fig. 3. Comparison of proposed options by evaluation indicators

4 Results and Discussion

Figure 2 shows the dependence of the change in the criteria for assessing measures when changing the intensity of vehicle traffic for different traffic management options.The analysis of the results of modelling the traffic management options according to the defined criteria at the intersection with changes in TF intensity showed the following:

- According to the criterion "average vehicle delay", at the base (968 vehicles/h) and reduced by 20% (774 vehicles/h) intensity, the engineering measure BSCs showed the best results. Providing 8.2 s and 5.5 s of average delay. Respectively. However, when the intensity was increased by 20% (1161 vehicles/h), the advantage in this criterion was observed in the traffic light control option. 11 s of average delay;
- According to the criterion 'number of vehicle stops', at all considered intensity levels. The best results are demonstrated by traffic light control. With the corresponding values of 0.61 (baseline intensity). 0.52 (–20%). And 0.58 (+20%) of the number of stops per vehicle;

– According to the criterion of 'average traffic jam length'. Traffic light control was the most effective at the basic (2.9 m) and increased (3.0 m) intensities. Instead, under the lowest load (774 vehicles per hour), the advantage remained with the BSCs, which provided the lowest value of the congestion. 1.7 m.

Thus, the modelling results confirmed the feasibility of introducing traffic lights at the intersection of A. Oshchepkova and Brigady Khartia streets as an effective means of traffic management. At the same time, one should not ignore the possibility of combining it with passive traffic calming measures, in particular within the street and road network. Among the alternatives. it is proposed to introduce Berlin cushions. Which have a minimal impact on the traffic dynamics at the intersection. While helping to reduce the accident rate and the severity of the consequences of accidents.

It is also worth noting that the difference between the efficiency of traditional road humps and Berlin cushions does not exceed 5%. However, the latter have a significant advantage in cases where public or freight transport moves along the street, as they do not cause excessive delays for these categories of road users.

Figure 3 shows changes in estimated indicators depending on the intensity of the TF according to the proposed study options. Which indicates the feasibility of using a particular means of traffic calming.

Using the statistical analysis package Statgraphics Centurion, we found the dependencies of the average value of the length of the traffic jam, the average number of stops, and the average delay on the intensity of traffic and pedestrian flows in different scenarios (Table 2).

These dependencies describe real indicators with a reliability of 91–96%, which indicates the adequacy of the built models. It has been proved that the performance indicators of a SRN section - in particular, the average length of a traffic jam, the number of stops, and the delay per vehicle, are significantly affected by both the availability of TCDs and the level of traffic intensity: with the growth of the latter, there is a tendency to increase all the estimated parameters.

It has been established that with an increase in the zone of influence and geometric parameters of TCDs, the values of the evaluated indicators increase proportionally. For example, an EPC has smaller geometric dimensions than a complex that combines a standard pedestrian crossing with an RH on both sides, and, accordingly, has a smaller impact on the flow of vehicles, which has a positive effect on reducing delays and congestion.

Table 2. Dependences of the average value of the length of the traffic jam, the average delay, and the average number of stops on the intensity of transport and pedestrian flows in different scenarios

Scenario number	The resulting dependency	Correlation coefficient
1	$\overline{L_{zat}^{\text{var1}}} = 14,65 - 0,009 \cdot N_t^* - 0,069 \cdot N_p^{**} + 0,0001 \cdot N_t^2 +$ $+0,0002 \cdot N_t \cdot N_p + 0,000064 \cdot N_p^2$	0.963

(*continued*)

Table 2. (*continued*)

Scenario number	The resulting dependency	Correlation coefficient
	$\overline{T_{zup}^{var1}} = 40,35 - 0,2435 \cdot N_t - 0,0672 \cdot N_p + 0,0003N_t^2 + +0,0003 \cdot N_t \cdot N_p + 0,000021 \cdot N_p^2$	0.94
	$\overline{N_{zup}^{var1}} = 2,238 - 0,0135 \cdot N_t - 0,004N_p + 0,000013 \cdot N_t^2 + +0,000016 \cdot N_t \cdot N_p + 0,000018 \cdot N_p^2$	0.92
2	$\overline{L_{zat}^{var2}} = 21,6 - 0,0949N_t - 0,1001 \cdot N_p + 0,000098 \cdot N_t^2 + +0,0002 \cdot Nt \cdot Np + 0,000095 \cdot N_p^2$	0.96
	$\overline{T_{zup}^{var2}} = 35,41 - 0,0718N_t - 0,2083 \cdot N_p + 0,0000438 \cdot N_t^2 + +0,0003 \cdot N_t \cdot N_p + 0,0003 \cdot N_p^2$	0.95
	$\overline{N_{zup}^{var2}} = 1,95 - 0,004N_t - 0,0112 \cdot N_p + 0,0000025 \cdot N_t^2 + +0,000015 \cdot N_t \cdot N_p + 0,000015 \cdot N_p^2$	0.93
3	$\overline{L_{zat}^{var3}} = 14,65 - 0,0901N_t - 0,0685 \cdot N_p + 0,0001 \cdot N_t^2 + +0,00022 \cdot N_t \cdot N_p + 0,0000639 \cdot N_p^2$	0.94
	$\overline{T_{zup}^{var3}} = 41,7 - 0,0861 \cdot N_t - 0,244 \cdot N_p + 0,000056 \cdot N_t^2 + +0,0003 \cdot N_t \cdot N_p + 0,0003 \cdot N_p^2$	0.95
	$\overline{N_{zup}^{var3}} = 1,68 - 0,005N_t - 0,0094 \cdot N_p + 0,0000043 \cdot N_t^2 + +0,000017 \cdot N_t \cdot N_p + 0,000013 \cdot N_p^2$	0.94
4	$\overline{L_{zat}^{var3}} = 21,5 - 0,0952N_t - 0,096 \cdot N_p + 0,000094 \cdot N_t^2 + +0,0002 \cdot N_t \cdot N_p + 0,0000949 \cdot N_p^2$	0.92
	$\overline{T_{zup}^{var3}} = 34,97 - 0,0712 \cdot N_t - 0,2063 \cdot N_p + 0,000044 \cdot N_t^2 + +0,0002 \cdot N_t \cdot N_p + 0,0003 \cdot N_p^2$	0.94
	$\overline{N_{zup}^{var3}} = 1,96 - 0,004N_t - 0,0113 \cdot N_p + 0,0000022 \cdot N_t^2 + +0,000015 \cdot N_t \cdot N_p + 0,000015 \cdot N_p^2$	0.93

Notes: *N_t intensity of the TF; **N_p - intensity of the pedestrian flow

The BSC is a local traffic calming measure that is particularly effective in car-dominated environments. Due to its design, it encourages passenger cars to slow down before pedestrian crossings, which reduces the risk of pedestrian collisions. At the same time, the BSC has minimal impact on large vehicles (buses, trucks, special vehicles), which can cross it without significantly reducing speed, which is an advantage for maintaining capacity. Thus, compared to a standard road hump, the BSC is a more flexible tool for ensuring pedestrian safety without causing excessive delays in traffic.

5 Conclusions

The study confirmed the effectiveness and adequacy of microscopic traffic modelling (PTV Vissim) in assessing the impact of various traffic calming measures under different traffic and pedestrian flow scenarios. In the absence of direct field data, internal validation based on the consistency of modelling results with theoretical expectations proved to be a reliable and scientifically grounded approach.

Figures 2 and 3 illustrate that the implemented hypothetical traffic calming scenarios led to a predictable and logical increase in key negative traffic performance indicators—such as delay, queue length, and number of stops—depending on traffic intensity. These trends align with established theoretical and empirical knowledge on the functioning of traffic systems under varying operational conditions.

Statistical analysis of the simulation results supports the robustness of the models. High correlation coefficients (0.88–0.99) across all regression equations indicate strong linear relationships. The Durbin–Watson statistics ranging from 2.37 to 3.56 with p-values > 0.1 confirm the absence of autocorrelation, while mean absolute errors between 0.02 and 1.17 demonstrate satisfactory predictive accuracy. The most accurate results were obtained in EI (Environmental Improvement) and EPC (Eco-Pedestrian Combined) scenarios (MAE (Mean Absolute Error) = 0.02–0.06), highlighting their potential as efficient traffic management strategies.

Table 2 further reinforces the models' validity by showing that across all four scenarios, the dependencies of delay, queue length, and number of stops on the intensities of vehicle and pedestrian flows exhibit strong correlations (0.92–0.96), confirming the models' sensitivity and responsiveness to real-world variables.

The comparative analysis demonstrates that the choice and configuration of traffic calming strategies can significantly influence traffic performance and environmental impact. Therefore, a data-driven and simulation-supported approach is crucial for selecting effective urban traffic management solutions to improve road safety, operational efficiency, and reduce emissions.

The developed simulation model can be effectively used as a decision-support tool for urban traffic planning authorities, enabling pre-implementation assessment of alternative designs and traffic control strategies. Future work should focus on expanding the model with real-world calibration data, integrating pedestrian behavior in greater detail, and analyzing the long-term effects of calming measures on modal shift, emission levels, and sustainable urban mobility.

References

1. Litman, T.: Traffic Calming: Benefits, Costs and Equity Impacts. Victoria Transport Policy Institute. Victoria, Canada (1999). www.vtpi.org

2. Road Safety. Traffic Calming Measures. General Technical Requirements. DSTU 4123:2020. [Effective from 2020-08-06], 43 p. https://surl.li/vwcqko
3. Buhaiov, I.S., Kholodova, O.O., Buhaiova. M.O.: Assessment of the effectiveness of traffic calming measures at an intersection. Mod. Technol. Mech. Eng. Transp. Lutsk **1**(20), 78–85 (2023). https://doi.org/10.36910/automash.v1i20.1036
4. Green, D., Lewis, K., Head, A.-M., Ward, J., Munro, C.: Guide to Traffic Management. Part 8: Local Street Management. Edition 3.0. Austroads Ltd. Sydney, Australia, 191 p. (2020). ISBN 978-1-925854-82-4. www.austroads.com.au
5. Bugayov, I., Kholodova, O., Buhaiova, M., Kholodov, O.: Case study on implementing traffic calming devices: an example from the street and road network in Kharkiv. Збірник матеріалів конференції **1**(1), 15–22 (2024). https://doi.org/10.36910/conf_avto.v1i1.1389
6. Berloco, N., Coropulis, S., Intini, P., Ranieri, V.: Effects of Berlin speed cushions in urban restricted speed zones: a case study in Bari Italy. Transp. Res. Proc. **60**, 180–187 (2022). https://doi.org/10.1016/j.trpro.2021.12.024
7. Berloco, N., Coropulis, S., Garofalo, G., Intini, P., Ranieri, V.: Analysis of the factors influencing speed cushion effectiveness in the urban context: a case study experiment in the city of Bari, Italy. Sustainability **15**(8), 6352 (2023). https://doi.org/10.3390/su15086352
8. Layfield, R.E., Parry, D.I.: Traffic calming – speed cushion schemes: prepared for Driver Information and Traffic Management Division. Department of the Environment. Transport and the Regions. Crowthorne: Transport Research Laboratory IV, 51 p. (1998)
9. Chang, K.N., Nolan, M., Nihan, N.L.: Developing design standards for speed cushions. Transp. Res. Rec. **2030**(1), 22–28 (2007). https://doi.org/10.3141/2030-04
10. Berthod, C., Leclerc, C.: Traffic calming in Québec: speed humps and speed cushions. J. Civ. Eng. Architect. **7**(4) (2013). https://doi.org/10.17265/1934-7359/2013.04.009
11. Cantisani, G., Corazza, M.V., Di Mascio, P., Moretti, L.: Eight traffic calming "easy pieces" to shape the everyday pedestrian realm. Sustainability **15**(10), 7880 (2023). https://doi.org/10.3390/su15107880
12. Jateikienė, L., Andriejauskas, T., Lingytė, I., Jasiūnienė, V.: Impact assessment of speed calming measures on road safety. Transp. Res. Proc. **14**, 4228–4236 (2016). https://doi.org/10.1016/j.trpro.2016.05.394
13. Gonzalo-Orden, H., Rojo, M., Pérez-Acebo, H., Linares, A.: Traffic calming measures and their effect on the variation of speed. Transp. Res. Proc. **18**, 349–356 (2016). https://doi.org/10.1016/j.trpro.2016.12.047
14. Gonzalo-Orden, H., Pérez-Acebo, H., Linares, A., Arce, M.: Effects of traffic calming measures in different urban areas. Transp. Res. Proc. **33**, 83–90 (2018). https://doi.org/10.1016/j.trpro.2018.10.079
15. Institute of Transportation Engineers. Traffic Calming Measures. https://www.ite.org/technical-resources/traffic-calming/traffic-calming-measures/. Accessed 10 January 2025
16. Traffic calming. https://surl.li/tqqqie. Accessed 10 January 2025
17. City of Winston-Salem. Traffic calming policy, May 2005. www.ci.winstonsalem.nc.us/DOT/trafficcalming.html. Accessed 6 September 2023
18. Younes, H., et al.: The traffic calming effect of delineated bicycle lanes. J. Urban Mob. 100071, June 2024. https://doi.org/10.1016/j.urbmob.2024.100071
19. Chng, S., Chang, C., Mosquera, K., Leong, W.Y.: Living in a silver zone: residents' perceptions of area-wide traffic calming measures in Singapore. Transp. Res. Interdisciplinary Perspect. 100710, December 2022. https://doi.org/10.1016/j.trip.2022.100710
20. Zargiannaki, E., Tzouras, P.G., Antoniou, E., Karolemeas, C., Kepaptsoglou, K.: Assessing the impacts of traffic calming at network level: a multimodal agent-based simulation. J. Traff. Transp. Eng. 41–54, February 2024. https://doi.org/10.1016/j.jtte.2023.01.003

21. Polloni, S.: Traffic calming and neighborhood livability: evidence from housing prices in Portland. Reg. Sci. Urban Econ. 18–37 (2019). https://doi.org/10.1016/j.regsciurbeco.2018.11.004
22. Juhász, M., Koren, C.: Getting an insight into the effects of traffic calming measures on road safety. Transp. Res. Proc. 3811–3820 (2016). https://doi.org/10.1016/j.trpro.2016.05.466
23. Advani, N., Danak, H., Dutta, M., Jena, S.: Spherical cap studs: a novel speed bump alternative to reduce discomfort with effective speed reduction. Traff. Injury Prevent. 228–236 (2024). https://doi.org/10.1080/15389588.2023.2278415
24. Doomah, M.Z., Paupoo, D.P.: Evaluation of the effectiveness of speed tables combined with other traffic calming measures and their community acceptance in Mauritius. Case Stud. Transp. Policy 1550–1565 (2022). https://doi.org/10.1016/j.cstp.2022.04.022
25. Rahman, F., Kojima, A., Kubota, H.: Transportation investigation on North American traffic calming device selection practices. IATSS Res. 105–119 (2009). https://doi.org/10.1016/S0386-1112(14)60249-1
26. Ahn, K., Rakha, H.: A field evaluation case study of the environmental and energy impacts of traffic calming. Transp. Res. Part D Transp. Environ. 411–424 (2009). https://doi.org/10.1016/j.trd.2009.01.007
27. Antić, B., Pešić, D., Vujanić, M., Lipovac, K.: The influence of speed bumps heights to the decrease of the vehicle speed – Belgrade experience. Safety Sci. 303–312 (2013). https://doi.org/10.1016/j.ssci.2013.03.008
28. Cottrell, W.D., Kim, N., Martin, P.T., Perrin, H.J.: Effectiveness of traffic management in Salt Lake City, Utah. J. Safety Res. 27–41 (2006). https://doi.org/10.1016/j.jsr.2005.08.007
29. Yeo, J., Lee, J., Cho, J., Kim, D.-K., Jang, K.: Effects of speed humps on vehicle speed and pedestrian crashes in South Korea. J. Safety Res. 78–86 (2020). https://doi.org/10.1016/j.jsr.2020.08.003
30. Nagrebelna, L.P., Belenchuk, O.V., Kononenko, A.O.: Raised pedestrian crossings as a means of ensuring pedestrian safety. Sci. Prod. J. **2**(254), 47–50 (2018). https://doi.org/10.33868/0365-8392-2018-2-254-47-50
31. Arsenieva, N.O., Fomenko, G.R.: Practice of forming traffic calming zones in cities. Mun. Econ. Cities. Ser. Tech. Sci. Architect. **3**(170), 168–173 (2022)
32. Liubyi, Y.V., Levchenko, O.S., Syromiatnikova, M.S.: Analysis of the efficiency of using mini-roundabouts. Sci. Bull. Lutsk Nat. Tech. Univ. **61**, 110–117 (2018). http://nbuv.gov.ua/UJRN/Nn_2018_61_20
33. DBN V.2.3-4:2015. Transport Facilities. Roads. Part I: Design. Part II: Construction. [Valid from 2016-04-01], Kyiv, 104 p. (2015). https://dreamdim.ua/wp-content/uploads/2022/08/DBN-V_2_3-4_2015.pdf
34. DSTU 4123:2020. Road Traffic Safety. Traffic Calming Devices. General Technical Requirements. [Valid from 2020-11-01], State Enterprise "DerzhdorNDI", Kyiv, 35 p. (2020). http://uas.gov.ua. (official distributor – SE "UkrNDNC")
35. Theodoulou, G., Wolshon, B.: Alternative methods to increase the effectiveness of freeway contraflow evacuation. Transp. Res. Rec. **1865**(1), 48–56 (2004). https://doi.org/10.3141/1865-08

Analyzing and Evaluating Safety in Numbers for Cyclists in a Large Dataset

Andreas Leich[1], Ronald Nippold[1], Louis Calvin Touko Tcheumadjeu[1], and Peter Wagner[1,2](✉)

[1] Institute for Transportation Systems, German Aerospace Center (DLR), Berlin, Germany
{andreas.leich,ronald.nippold,louis.toukotcheumadjeu, peter.wagner}@dlr.de

[2] Institute for Land- and See Traffic, Technical University of Berlin, Berlin, Germany

Abstract. In traffic safety, the so-called safety-in-numbers (SiN) is a well-known effect that is often hard to pin down, and, especially in the not-so-scientific literature, overwrought with too much optimism. Since traffic safety and vision zero is an important building block of an intelligent transportation system, this paper is devoted to describe and analyze this effect with the help of two large databases from Germany. Not too surprisingly, overly optimistic claims like decreasing crash numbers of bicyclists for large bicycle numbers can be dismissed by this analysis, and a much more modest result comes forward: the number of bicycle crashes grows sub-linear with the number of bicycles.

Keywords: Traffic safety · bicycle · safety-in-numbers · smart data

1 Introduction

The Safety in Numbers (SiN) effect states: if there is a lot of bicycle or pedestrian traffic $Q_{\text{cyc,ped}}$ on a road, then this road becomes safer for bicyclists or pedestrians, respectively. This safety is measured in terms of the number of crashes N_{cyc} in a certain time interval, but other values like SSM (safety surrogate measures) might be used as well. Different views are possible what safer means, in some cases people believe that the number of crashes does really decrease with the number of VRU (vulnerable road users).

For extreme cases, as depicted in Fig. 1 this might be true: if the bicycle share is very large, then no car can drive faster than the bicycles. For further reference, this effect will be named SiLN, Safety in Large Numbers. It is well different from the "normal" SiN, which will be defined below.

Except for SiLN, there is anecdotal evidence that in some cases, bicycle crash numbers have been decreasing with larger bicycle traffic, especially when followed over longer timescales [1, 2]. We do not think this to be true, and we will use the opportunity of two large data sets to correct this view.

In traffic safety research, SiN is associated with a sub-linear growth of the number of bicycle crashes with the number of bicycles on the road. See [3–7], and [8] for a recent

E. Macioszek et al. (Eds.): TSTP 2025, LNNS 1769, pp. 75–85, 2026.
https://doi.org/10.1007/978-3-032-14094-4_6

literature review. It claims that:

$$N_{\mathrm{cyc}} \propto Q_{\mathrm{cyc}}^{\beta} \tag{1}$$

where the exponent $\beta < 1$ represents this sublinear growth. This is also not the strongest argument either, since $\beta < 1$ has been observed in almost all investigations in which the influence of the other crash partner has been left out, which is even true for cars [9]. However, this discussion is not in the focus of this paper.

Fig. 1. An example of a (very) large bicycle flow at an intersection. Own picture.

Here, the following two datasets have been utilized (another large-scale approach is reported in [8]):

- The German Accident Atlas (GAA) 2016–2024 [10], which samples crashes with injured and killed participants,
- bicycle count data from the company eco-counter (ECC – eco counter cycling) over the years 2016–2024.

Note that both datasets are open: the GAA data can be downloaded directly, while the ECC [11] data can be obtained through Eco-counter's API [12]. The Fig. 2 displays this process.

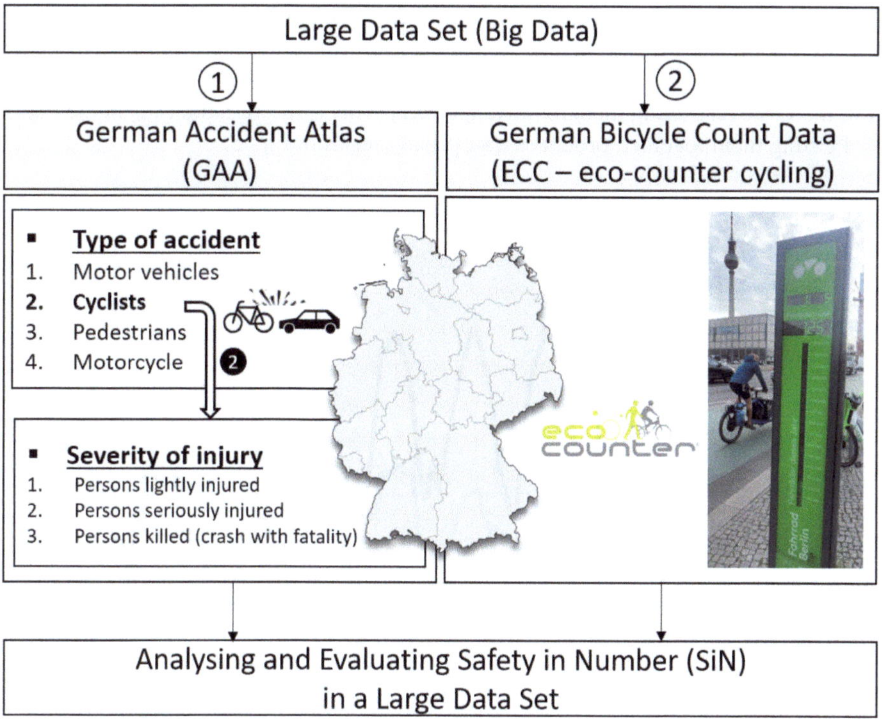

Fig. 2. Data sets used in this contribution.

2 Research Methodology

2.1 Reading the Data

A reasonable idea when it comes to data cleaning is to leave data as raw as possible, since any tweaking may reflect prejudices of the data scientist. Therefore, only a minimum amount of cleaning has been performed: four ECC detectors have been discarded, since they had negative counts. Note, too, that the number of counters has changed considerably over the years, from 147 in January 2016 to 802 in August 2024, see Fig. 4 (right). Also, some counters vanished from the record during these seven years, since the total number of different ids in this dataset is 997.

Therefore, we obtain datasets where flow data and crash data for bicycles from a whole country can be correlated. The ECC data are monthly averages for each detector, the GAA data may have a finer time resolution but got aggregated to monthly data as well. In addition, they must be scaled so that they can be compared to each other.

2.2 Data Fusion

When fusing the GAA with the ECC data, two challenges are encountered: GAA data are incomplete before 2020. This is due to the slow build-up of the GAA data-base, in the beginning only half of the German federal states have been present. Therefore, the share of (monthly) bicycle crashes in all monthly crashes is used here, instead of the raw crash numbers, which is named p_{cyc} in the following and in Fig. 3. The eco counter flows have been normalized as well, here they have been divided by their maximum flow, therefore restricting them to the range [0,1]. This is the variable $\tilde{Q}_{\mathrm{cyc}}$ in the Fig. 3.

Plotting them together, produces a surprising result in Fig. 3.

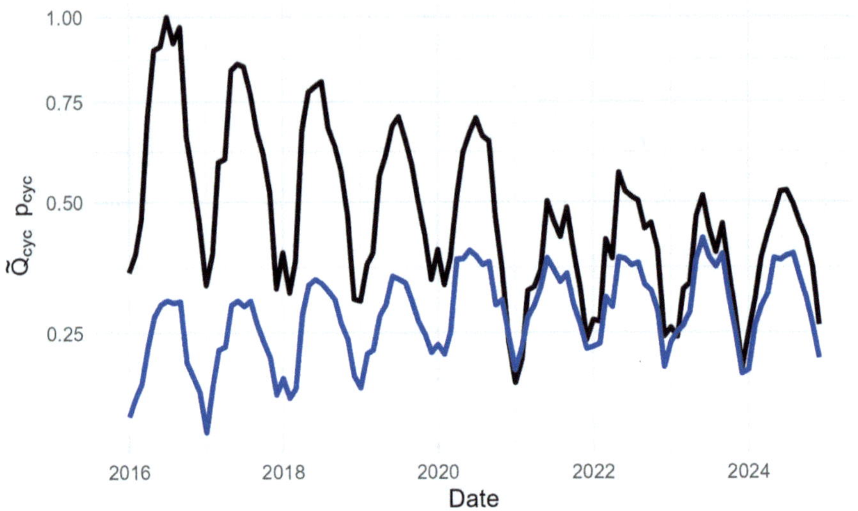

Fig. 3. Normalized number of bicycle counts $\tilde{Q}_{\mathrm{cyc}}$ from the ECC (black line), plotted together with the share of bicycle crashes p_{cyc} (blue line).

The apparent observation in Fig. 3 is that the two curves are very strongly correlated, but also very clearly follow a different trend. After de-trending, the correlation between them reaches a very good correlation of 0.87, without de-trending, it is 0.41. The de-trending has been performed only for the purpose to compute the correlation; it has been done by subtracting a linear fit from each of the two curves.

This is one example where it is almost certain that this result is wrong. This is because other data from travel surveys have recorded an increase in the bicycle share of traffic in Germany during the last 10 years. Looking deeper, two reasons come to mind. Both have been already mentioned: the crash data cover the whole of Germany only since 2020; however, we tried to correct this by computing the relative index, i.e. the share of bicycle crashes of all crashes. For the ECC, there is the problem that its numbers have changed strongly during the study period. Both observations are shown in Fig. 4. One may speculate that this introduces a bias, since cities may install bicycle counters first at places with large bicycle counts, leaving smaller flow sites for the later installations.

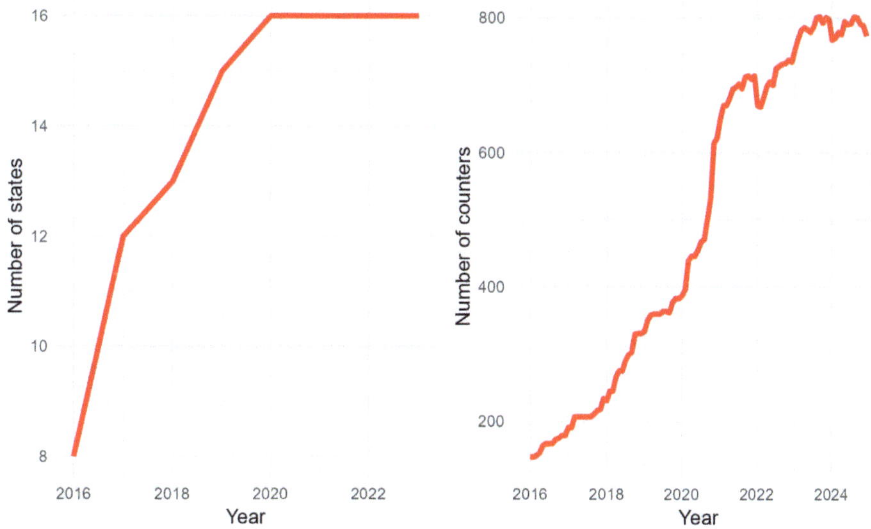

Fig. 4. Growth of the number of Federal States contributing to GAA (left), and growth of the number of Eco-Counter detectors (right).

Figure 5 depicts the mean bicycle count as function of first installment month of the respective detector. Note, there is still a bias in this data: only complete years should enter into the computation of the mean value.

To circumvent the effect of the installment time, each detector is analysed individually as follows: in a first step, each detector time-series is normalized to its 95-percentile of the monthly counts. In addition, each detector gets assigned two quality values: the first one measures the percentage the detector has a monthly value in comparison with the maximum possible maximum number of monthly counts. The second quality value is the ratio between the maximum count and the 95-percentile. Only data are used where more than 90% of monthly data are available, and where the ratio between the maximum and the 95-percentile was below 1.5. This corrects mostly for the installation times of the detectors, and for outliers. In addition, the scaled flow from the eco counter have been used subsequently in the plots and analyses. Therefore, a new variable name has been introduced for this normalized value, it is named $\widehat{Q}_{\mathrm{cyc}}$.

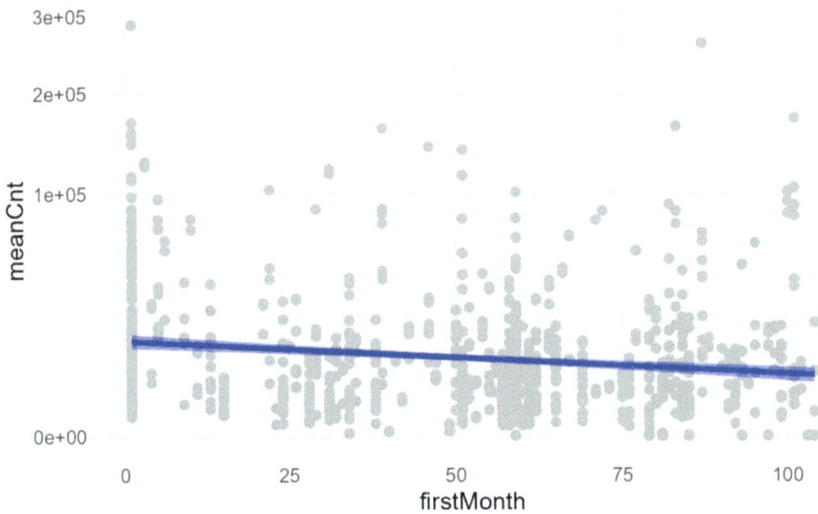

Fig. 5. Mean bicycle count as function of first installment month of the respective detectors. The line is a linear fit to the data, clearly indicating the decreasing count with the month of installment.

3 Results Presentation

After the pre-processing described above, the data have been analyzed as follows. In a first step, the presentation in Fig. 3 is repeated, and now a result that is in line with travel demand surveys in Germany is obtained. The results are displayed in Fig. 6.

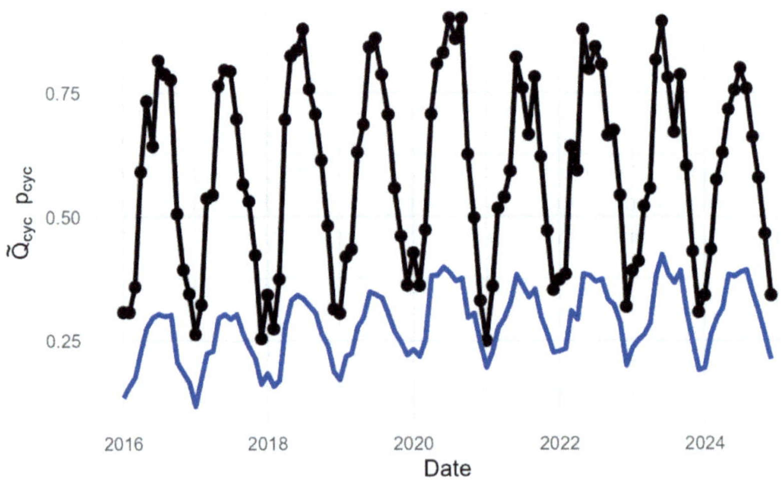

Fig. 6. The monthly scaled values of the number of crashes (blue) and the bicycle traffic in Germany (black). Note the strong synchronicity between the two curves

3.1 Different Representations

These data can be represented in a more common format, i.e. the number of crashes in relation to the number of bicycles, both in the normalized form $\hat{N}_{cyc} := p_{cyc}$ versus $\hat{Q}_{cyc}$, which is shown in Fig. 7.

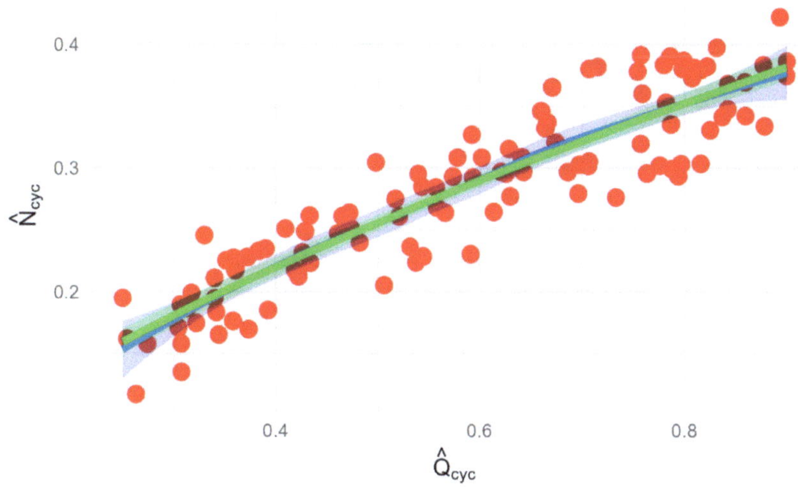

Fig. 7. Scaled number of crashes versus the scaled number of bicycles, together with a GLM-fit (green) to Eq. (1) and a local smoothing spline (blue).

Models can be fitted to these data. In a first step, a GLM with the power-law $\propto \widehat{Q}^{\beta}$ is fitted. This differs a bit from what is usually done in traffic safety, since the crash count variable is continuous, not discrete as in most traffic safety analyses. Therefore, a normal distribution is used, together with a logarithmic link function. Alternatively, a Gamma distribution, also with a logarithmic link function, has been used and yielded a slightly smaller AIC, but the same value of β. So, it turns out that based on these data, the number of bicycle crashes in Germany grows with the number of bicycles observed sub-linearly, with an exponent $\beta = 0.68 \pm 0.03$. The error probability when testing this hypothesis is very small ($p < 10^{-16}$).

In a last step, the crash data are disaggregated one step further. The GAA has for each crash the "category", which is the severity of the crash:

1. corresponds to persons lightly injured
2. to persons seriously injured
3. to persons killed (for a crash with fatalities)

The same treatment as above can now be performed on the three severities. This leads to the results presented in Fig. 8. By analyzing them in the same manner, almost the same power-law for all three severity grades could be observed. All this manifests a robust SiN effect on the scale of a whole country, but there are no immediate surprises.

The exponent itself is in the range found in other investigations, which are typically much smaller study areas.

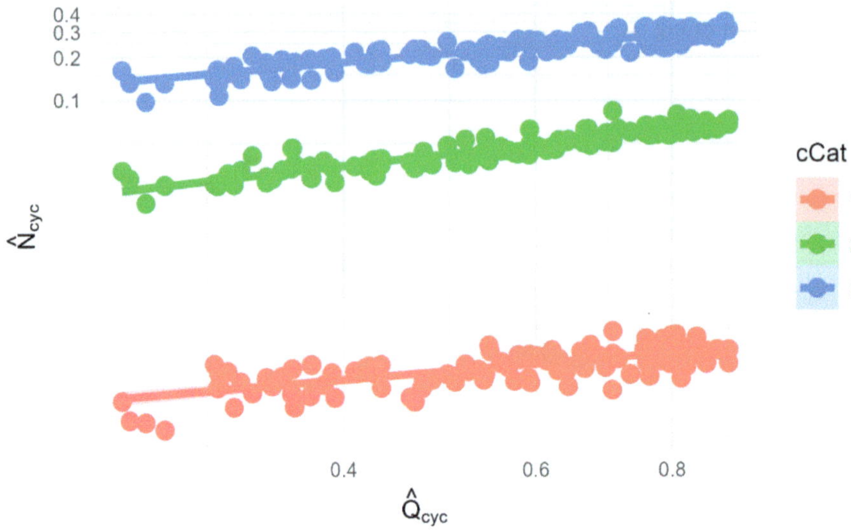

Fig. 8. Scaled number of crashes versus the scaled number of bicycles for the three degrees of crash severity in the data, together with a GLM-fit (line) to Eq. (1). The exponents from the fits are $\beta_{1,2,3} = 0.62, 0.84, 0.65$. The y-axis has been scaled by a sqrt to improve readability

4 Considerations About Modeling

There are several ways to make progress here: one could investigate more detailed datasets, or one can try to model the whole process of crash generation. Here, we follow the modeling approach. The causation of crashes is arguably a complicated process, here we will look only at the relationship between car and bicycle flow and crashes. In our view, this is an important determinant of crashes, on par with the most important infrastructure features. We keep in mind that the empirical results on the most aggregated level of flows such as used here (or, in other work, the AADT = average annual daily traffic is used instead) show that the number of bicycle crashes grows with the number of bicycles and cars sub-linearly. This finding should come out of any more detailed microscopic model.

In the following it is assumed that this sub-linear growth has a certain mechanism behind. For instance, when we look at the interaction between car drivers and a group of bicyclists with n_{cyc} bicycles, then one can assume that for any interaction, there is a small probability p_0 that the car driver is overlooking the bicyclists. To have a SiN effect, this can be modelled as:

$$p_{\text{err}} \propto p_0^{n_{\text{cyc}}}, \qquad (2)$$

Here, the probability of over-looking decreases exponentially with the number of bicycles present. Another approach was motivated by the work of Ekman [13]. From this work, a kind of step-function can be assumed, which states a two-stage error rate: for small bicycle numbers n_{cyc}, the error probability is given by some p_0, which decreases to a lower level for a larger exposition of the bicycles, named $p_\infty < p_0$:

$$p_{\text{err}} \propto \begin{cases} p_0 & n_{\text{cyc}} < n_{\text{crit}} \\ p_\infty & \text{else} \end{cases}, \tag{3}$$

These microscopic models can be extrapolated to a macroscopic level as follows: we have used an open source microscopic traffic flow simulator (SUMO, [14]) to simulate an intersection, where bicycle streams crosses vehicle streams, both of varying strengths (expositions) q_{cyc} and q_{veh}, see Fig. 9.

Then, we can count the vehicles in the two boxes for a small time interval (30 s aggregation have been used here) and apply the two microscopic error models, Eqs. (2) or (3). This yields a probability that in this time interval a crash will occur. These probabilities can then be summed and sorted according to the expositions. Interestingly, this in a SiLN, but not a SiN. If the microscopic assumptions above were true, and if the general approach sketched here is valid as well, then the crash-rate of bicycles would actually increase with the number of bicycles, which contradicts the empirical findings in this work, but also in other investigations.

So far, we have not found a microscopic law that leads to the empirically observed power-law in Eq. (1), this is still under investigation.

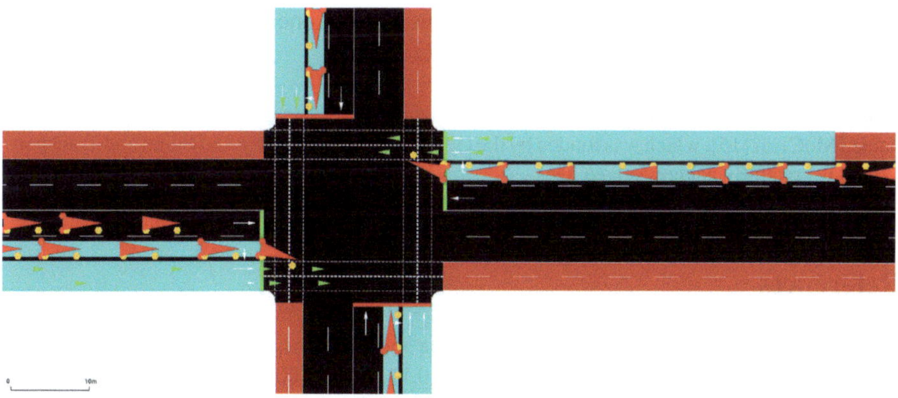

Fig. 9. Screenshot of a SUMO simulation, showing cars (red) and bicycles (green), and the areas where the number of objects (n_{cyc}, n_{car}) are counted in cyan

5 Conclusions

On this level of the aggregation of bicycle traffic flow one can rule out that the number of injured bicyclists is decreasing with the number of bicycles traveling on Germany's roads. Instead, the total number grows sub-linearly in the number of bicycles, so in the sense that SiN is interpreted in traffic safety research, a very robust SiN effect is indeed visible.

We have also made first steps into a microscopic modeling of the crashes between cars and bicycles. We think that this is a promising approach to be pursued further: first, the interactions between a single driver and a group of bicyclists might be accessible to driving simulators or even video-based observations of the interaction between cars and groups of bicyclists. And second, a certain number of error models like Eq. (2) or (3) could be tested to lead to the observed sub-linear growth in Eq. (1). Of course, they should finally be in line with the results from the observations above.

As a last point, we want to pick up on one remark from the introduction. SiN is related to an exponent smaller than one in the crash-number-exposition relationship. However, especially Eq. (1) states just one half of the truth. In reality, an extended model reads (and is in fact used in many studies):

$$N_{\text{cyc}} \propto Q_{\text{cyc}}^{\beta_{\text{cyc}}} Q_{\text{veh}}^{\beta_{\text{veh}}}. \tag{4}$$

Quite clearly, the number bicycle crashes should depend on the number of vehicles present. The modeling approach of Sect. 4 used such a dependency. For the data here, we do not have Q_{veh}, so we ignored its effect. This is done in some studies on bicycle or pedestrian crashes as well, so that some researchers even claim that all exponents have to be smaller than one.

The meta-study [5] states that for multi-vehicle crashes of two vehicles ($N_{\text{veh,veh}} \propto Q_{\text{veh}}^{\beta} Q_{\text{veh}}^{\beta} = Q_{veh}^{2\beta}$), the exponent 2β is between 1 and 1.4 or so, leading to $\beta \in [0.5, 0.7]$. Which is in line with the exponent found here. This, however, states that there is a SiN even for vehicles. While this may not come unexpected, one may question that the two SiN's (for vehicles and bicycles) are comparable.

Acknowledgment. We learned how to deal with Eco-Counter data in a German Research Project named MoCKiii [15], which was funded by the German Federal Ministry for Digital and Transport. Which is gratefully acknowledged here.

References

1. Jacobsen, P.: Safety in numbers: more walkers and bicyclists, safer walking and bicycling. Injury Prevent. **9**, 205 (2003). https://doi.org/10.1136/ip.9.3.205
2. Wikipedia contributors, Safety in Numbers - Wikipedia, the Free Encyclopedia (2024)
3. Elvik, R.: The non-linearity of risk and the promotion of environmentally sustainable transport. Accid. Anal. Prev. **41**, 849 (2009). https://doi.org/10.1016/j.aap.2009.04.009
4. Elvik, R.: Can a safety-in-numbers effect and a hazard-in-numbers effect co-exist in the same data? Accid. Anal. Prev. **60**, 57 (2013). https://doi.org/10.1016/j.aap.2013.08.010
5. Elvik, R., Goel, R.: Safety-in-numbers: an updated meta-analysis of estimates. Accid. Anal. Prev. **129**, 136–147 (2019). https://doi.org/10.1016/j.aap.2019.05.019
6. Elvik, R., Bjørnskau, T.: Safety-in-numbers: a systematic review and meta-analysis of evidence'. Saf. Sci. **92**, 274–282 (2017)
7. Kehoe, N.P., Goughnour, E., Jackson, S., Sykes, K., Miller, S., Blackburn, L.: Safety in numbers: a literature review (Report No. DOT HS 813 279). National Highway Traffic Safety Administration (2022)

8. Aldred, R., Goel, R., Woodcock, J., Goodman, A.: Contextualising safety in numbers: a longitudinal investigation into change in cycling safety in Britain, 1991–2001 and 2001–2011. Inj. Prev. **25**, 236–241 (2019). https://doi.org/10.1136/injuryprev-2017-042498
9. Høye, A.K., Hesjevoll, I.S.: Traffic volume and crashes and how crash and road characteristics affect their relationship, a meta-analysis. Accid. Anal. Prev. **145**, 105668 (2020). https://doi.org/10.1016/j.aap.2020.105668
10. Statistical Administration of the German Federal Government and the Federal States, German Accident Atlas (2025). https://unfallatlas.statistikportal.de/
11. Eco Vision, Eco Counter Webpage (2025). https://www.eco-counter.com
12. Eco Vision, API Documentation (2025). https://github.com/bundesAPI/eco-visio-api
13. Ekman, L.: On the Treatment of Flow in Traffic Safety Analysis, - a non-parametric approach applied on vulnerable road users, Doctoral Thesis (monograph), Transport and Roads (1996)
14. Lopez, P.A., et al.: Microscopic traffic simulation using SUMO. In: 2018 21st International Conference on Intelligent Transportation Systems (ITSC), Maui, HI, 04.11.2018–07.11.2018, pp. 2575–2582. IEEE (2018)
15. MoCKiii, ITS German research project webpage (2025). https://www.stadt-koeln.de/artikel/72396/index.html

Smart, and Ecological Solutions

A Microsimulation-Driven Approach to Assess Performance of Alternative Intersection Designs Embedded with Smart Vehicle Technologies

Anna Granà[1], Elżbieta Macioszek[2], and Maria Luisa Tumminello[1](✉)

[1] Department of Engineering, University of Palermo, Palermo, Italy
{anna.grana,marialuisa.tumminello01}@unipa.it

[2] Department of Transport Systems, Traffic Engineering and Logistics, Faculty of Transport and Aviation Engineering, Silesian University of Technology, Katowice, Poland
Elzbieta.Macioszek@polsl.pl

Abstract. The paper presents a framework driven by microsimulation to assess the performance of connected automated vehicles at intersection designs. The research explores the behaviour of connected automated vehicles in mixed traffic, focusing on calibrating model parameters to align simulated data with entry capacity functions for various deployment rates of smart vehicles. A suburban signalized intersection in Poland serves as the baseline case study for modelling the transition to a knee-turbo variant in Aimsun. The methodology employs a validation-focused approach that highlights parallels in entry mechanisms at the tested roundabouts. Findings reveal that the performance of connected automated vehicles is significantly affected by intersection geometry and the deployment rates of these vehicles, providing insights for infrastructure project assessments and support the integration of road designs with smart driving technologies.

Keyword: road design signalized intersection roundabout smart mobility microsimulation sustainability Aimsun

1 Introduction

As the demand for sustainable transportation systems grows, skilled road engineers are essential for developing innovative solutions that effectively address mobility challenges through the integration of emerging technologies, road design, industrial innovation, performance optimization, and circular construction practices [1]. Connected automated vehicles (CAVs) are expected to enhance citizen well-being by reducing crashes, increasing road capacity, lowering emissions, and improving traffic management through advanced communication technology in smart vehicles [2–4]. Challenges in the practical implementation of CAVs remain, especially since high levels of automation are not yet available on roads; however, cooperative adaptive cruise control can adjust speeds and distances, allowing CAV platoons to reduce delays [5–7]. Nodes in the road network, such as intersections and roundabouts, are critical for traffic management, significantly influencing traffic efficiency, crash prevention, and ensuring a smooth flow of vehicles

E. Macioszek et al. (Eds.): TSTP 2025, LNNS 1769, pp. 89–105, 2026.
https://doi.org/10.1007/978-3-032-14094-4_7

[8, 9]. Cautious human drivers complicate negotiations at intersections, while signalized intersections simplify gap acceptance [10]. Roundabouts are cost-effective, but pose challenges for CAVs, particularly regarding curved trajectories and lane changes [11, 12]. In turn, turbo roundabouts effectively use curbs to separate traffic flows and aid navigation [13]. Studies on CAVs increasingly use micro-simulation models to analyse their performance in mixed traffic, focusing on interactions with driver-operated vehicles (DoVs) and evaluating algorithm effectiveness [12, 14, 15]. Research shows smart vehicles surpass traditional vehicles in capacity due to advanced connectivity and automation, despite differing communication levels [16, 17]. Current studies are also exploring trajectory design for CAVs and DoVs to enhance performance at signalized intersections, using machine learning [18]. Effective navigation rules and optimal vehicle path management are also crucial for roundabouts [19]. However, machine learning techniques involve high computational costs and lengthy processing times. Micro-simulation results provide valuable insights, but diverse case evaluations are needed to improve generalizability across studies [14]. Despite advancements in CAV technology, comparative analyses of different geometric design proposals at intersections are still lacking. This study seeks to address this gap by evaluating various intersection design alternatives and the impact on the operational efficiency of smart vehicles using microsimulation.

This paper explores critical questions related to Aimsun's parameter calibration, assessing its accuracy in replicating simulated data against established entry capacity functions (ECFs) under varying CAV deployment rates (CAV-DRs) [20]. It evaluates how different geometric and functional designs of intersections influence operational performance as CAV-DRs shift. Also, it examines the feasibility of incorporating performance criteria to assess various geometric designs using entry mechanism analogies. The study uses a Polish signalized intersection analysed in prior research as a baseline for modelling the transition to a knee turbo variant within a microsimulation framework [21]. This also includes a two-lane roundabout design and various traffic scenarios with differing CAV-DRs. The study introduces a methodological framework for evaluating geometric and functional designs while assessing the impacts of changing intersection schemes on CAV performance. Aimsun enables the authors to assess the alignment of simulated outputs with ECFs, factoring in driving parameters and correction factors. By paralleling entry mechanisms between roundabouts, a validation-driven approach is tested. The findings aim to improve the design process and support decision-making for alternative project proposals, facilitating the evaluation of smart technologies' anticipated benefits in a roadway context.

After the introduction, the paper presents the methodology in Sect. 2, which outlines the intersection design alternatives and Aimsun modelling. Section 3 presents and discusses the results, while Sect. 4 concludes the paper.

2 Methodology

The methodological path began with the selection of the baseline case study. A suburban signalized intersection in Poland, which had previously been used in research [21], was identified based on the size and organization of turning movements. This intersection now serves as a basis for modelling the transition to a turbo variant in a microsimulation environment. Additionally, a two-lane roundabout was designed to represent an intermediate step in the evolution of the intersection design integrated with vehicle-to-vehicle communication technologies. Thus, the definition of different intersection designs can facilitate the operational assessment of nodes that are configured and organized differently at the same location within the road network. Subsequently, additional data on traffic patterns, vehicle types, and signal operations were gathered to complement the available information regarding origin-destination traffic and vehicle movements [21]. This provided the basis for designing a two-lane roundabout [9], aiming to transform it into a knee turbo roundabout [13, 22]. Traffic scenarios were established with CAV-DRs from 0% to 100% in 20% increments. ECFs were then generated for each intersection, incorporating correction factors for CAVs [3]. In Aimsun, vehicle-to-vehicle communication for mixed fleets of CAVs and DoVs replicated traffic conditions for the proposed layouts, followed by sensitivity analysis and calibration based on key model parameters. Calibration accuracy for the signalized intersection was tested by adjusting green time values within the traffic light cycle at various CAV-DRs. The knee turbo roundabout validated the two-lane roundabout model, as their entry mechanisms are similar. The performance of CAVs across different layouts was compared using selected key performance indicators. Figure 1 shows the process map of the proposed methodology.

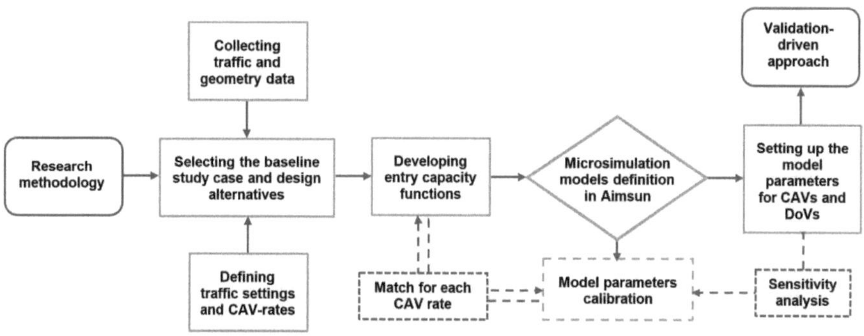

Fig. 1. The process map of the research methodology.

2.1 Intersection Design Alternatives

To evaluate the operational performance of mixed fleets of CAVs and DoVs at the intersection level, an isolated signalized intersection in Opole City, Poland, was chosen as the baseline case study. Figure 2 illustrates the intersection sketch along with the main results of the field data collection that took place during peak hours in September 2024. The signalized intersection comprises four approaches: Approaches 1, 3, and 4 in Fig. 2a feature one shared right-turn and through lane, plus an exclusive left-turn lane; Approach 2 in Fig. 2a has three entry lanes, each designated for a specific movement. The main road in the northwest and southeast directions includes two entry lanes and one exit lane; the minor road extends northeast with two entry lanes and one exit lane, and southwest with three entry lanes and one exit lane. Each entry lane measures 3.25 m; each exit lane measures 3.50 m. Signal cycle length, vehicle manoeuvrings, and composition data were videotaped from the approach 3 (see Fig. 2a). The average traffic volume was 2,263 vehicles per hour, with 90% cars, 9% heavy vehicles, 1% motorcycles and mopeds, and a low percentage of pedestrians and cyclists due to the suburban context of the installation.

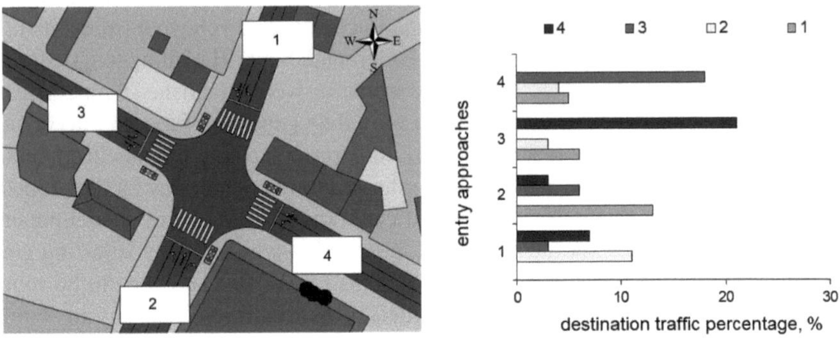

(a) (b)

Fig. 2. The baseline study case (50° 39′ 17″ N; 17° 57′ 16″ E): (a) geometric sketch; (b) bar graph of the origin-destination traffic where 1, 2, 3, and 4 represent the entry approaches.

Figure 2b shows the distribution of vehicle maneuvers: 57% of traffic travelled the main road (path 3–4 and vice versa), which included 39% through movements, 10% left turns, and 8% right turns, with a 3% difference in the directional distribution of traffic favouring the path 3–4. Conversely, 43% of traffic travelled the minor road (direction 1–2 and vice versa), including 24% through movements; only a 1% difference in the directional distribution of traffic favoured the path 2–1. The fixed four-phase timing schedule of the traffic light had a cycle duration of 120 s, with the distribution of green time as follows (refer to Fig. 2a for entry approaches): 27% for phase 1 assigned to right turns and through traffic from approaches 1 and 2; 10% for phase 2, assigned to left-turning vehicles from approaches 1 and 2; 43% for phase 3, assigned to right turns and through movements from approaches 3 and 4; 10% for left turns from approaches 3 and 4, as well as right turns from approach 2 in phase 4. An individual lane group was designated for exclusive left turns, while right turns and through movements shared a single lane group [3].

The size of the intersection area, suitable for a circular layout, along with its suburban context and traffic characteristics, led to the decision to convert the existing signalized intersection into a knee turbo roundabout. Initially, a compact, two-lane roundabout was implemented as a transitional measure. This interim step aimed to facilitate the eventual upgrade to a knee turbo roundabout. Figure 3 shows both circular schemes: approach 1 (minor road) and approaches 3 and 4 (main road) feature two entry lanes and one exit lane; approach 2 (minor road) features two entry lanes and two exit lanes on the roundabout, and one entry lane and two exit lanes on the turbo counterpart. The roundabout design in Fig. 3a includes a non-traversable central island, an inscribed circle radius of 13 m, two circulatory lanes, raised splitter islands, and deflection angles exceeding 45°. In contrast, the turbo roundabout design in Fig. 3b is inspired by Fortuijn [22] as an alternative layout aimed at accommodating travel demand within the intersection while respecting the surrounding built context. To ensure safe traffic management, splitter islands were installed [13, 23, 24]. Table 1 provides details about the geometry of both roundabouts.

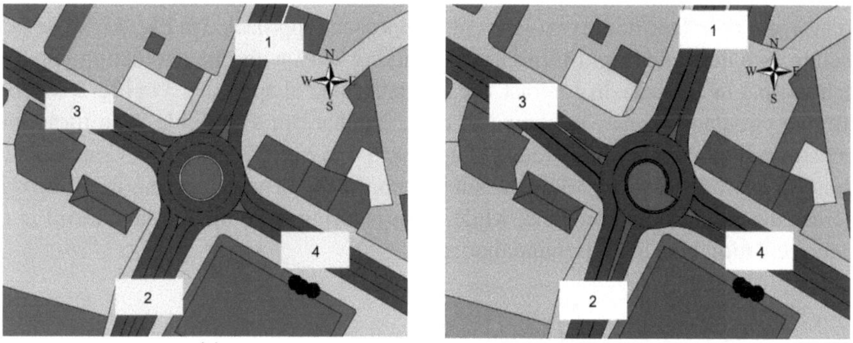

(a) (b)

Fig. 3. The conceptualised case studies: (a) Two-lane roundabout (Roundabout 1); (b) Knee turbo roundabout (Roundabout 2).

Table 1. Geometric design elements of roundabout and knee turbo roundabout.

Design element	Lane width [m]	
	Roundabout	Knee turbo roundabout
Circulatory roadway	3.75	4.00
Entry lane	3.75	3.75
Exit lane	3.50	3.50

The knee turbo roundabout design aims to validate the roundabout model by using an analogy in entry mechanisms. This analogy illustrates the similarities in how vehicles enter and navigate both standard roundabouts and turbo roundabouts based on the entry approach. An entry manoeuvring scheme for each approach outlines specific pathways for vehicle movements, detailing lane utilization and potential conflicts with circulating traffic. This scheme is helpful for understanding traffic dynamics, safety, and efficiency at entry points. Consequently, it supports hypotheses regarding traffic patterns by providing an analogy in entry mechanisms, thereby enhancing our understanding of vehicle behaviour in circular configurations. Table 2 outlines the potential entry mechanisms at roundabouts and turbo roundabouts, which depend on design characteristics, including the combinations of entry lanes and conflicting circulating lanes for each entry approach. However, the configuration of a roundabout or turbo roundabout—and consequently, the driving behaviour—can also vary based on the number of exit lanes and constraints imposed by the built or natural environments where the road infrastructure is installed or should be implemented. In both the knee turbo roundabout and the standard roundabout, drivers adhere to similar entry rules, yielding to circulating traffic and selecting appropriate lanes based on their intended direction. This consistency enhances the effectiveness of validating the simulated models, as vehicle behaviour is expected to align at both roundabouts due to their comparable geometries and traffic operations. For instance, the right lane of the two-lane entry approach (approach 3 in Fig. 3) can de facto function as a dedicated right-turn lane, conflicting only with one circulating lane (see mechanism 1 in Table 2). In contrast, the left entry lane of approach 3 in Fig. 3 conflicts with two circulating lanes, as shown in entry mechanism 4 in Table 2. In turn, entry mechanism 3 in Table 2 can be found at approach 4 in Fig. 3b and so on. Consequently, assuming approach 3 in Fig. 3 as the subject approach, drivers entering from the right lane use the outer circulating lane, while those entering from the left lane travel in the inner circulating lane at both roundabouts.

Table 2. Entry mechanisms by roundabout or turbo roundabout approach.

No	Manoeuvring	Approach[a]	Description
1	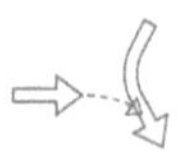		One entry lane conflicted with one circulatory lane, typical of the entry approach to a single-lane roundabout [9] and a turbo roundabout with a single circulatory lane [13][24] (e.g., Approach 2, Roundabout 2 in Fig. 3b).
2	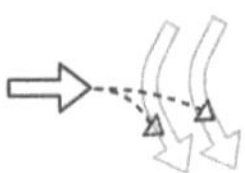	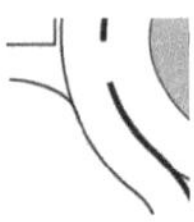	One entry lane conflicted with two circulating lanes: a single-lane entry for roundabouts or turbo roundabouts with two circulatory lanes. The driver enters the roundabout and decides which lane to use based on the desired destination [9][13][24,25]
3	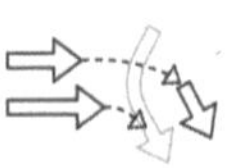	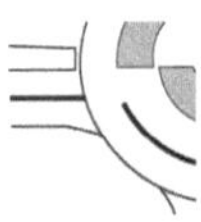	Two-lane entry conflicted with one circulatory lane for each entry lane for two-lane roundabouts [9] and turbo roundabouts with a single circulating lane where the inside circulatory lane is only added at one approach; e.g., knee turbo roundabouts (see Approach 4, Roundabout 2, Fig. 3b) or at two-lane entry approaches aligned in the same direction (e.g., basic or egg turbo roundabouts) [13][24].
4	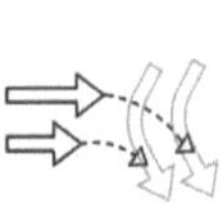	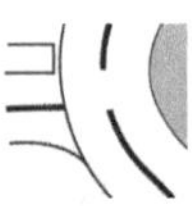	Two-lane entry conflicted with two circulatory lanes for each entry lane: drivers preselect the lane before entering and then move until the desired exit at two-lane or turbo roundabouts [9][13] (e.g., Roundabout 1 in Fig. 3a and two-lane entry approaches 1, 3 at Roundabout 2 in Fig. 3b).

[a] Roundabout or turbo roundabout entry approach.

2.2 Aimsun Modelling

In the research scenario, simulations began with a baseline of 100% DoVs and 0% CAVs. Subsequently, various traffic configurations were explored in Aimsun, increasing the percentage of CAVs by increments of 20%. The configurations examined included: a) 20% CAVs and 80% DoVs, b) 40% CAVs and 60% DoVs, c) 60% CAVs and 40% DoVs, d) 80% CAVs and 20% DoVs, and e) 100% CAVs and 0% DoVs. The study initiated with the development of micro-simulation models for the signalized intersection and two-lane roundabout. The calibration concentrated on these intersection models; the knee turbo roundabout served as a validation tool to assess calibration precision, using the analogy of yield negotiation at both roundabout entries in Fig. 3. Thus, validation employed the knee turbo roundabout as a control scheme to assess calibration accuracy. Approaches 3 in Fig. 3 were used as the subject approaches to present the analysis path.

Entry capacity functions (ECFs) were created for the signalized intersection and two-lane roundabout, addressing the absence of high-automation vehicles in traffic. Equations 1, 2, and 3 facilitated the development of these ECFs, enabling capacity calculations that account for varying CAV-DRs at the intersection level:

$$C_{CAVs} = \frac{G \cdot N_l \cdot S_{fr,CAVs}}{c} \tag{1}$$

where: C_{CAVs} is the entry capacity, pc/h, at the signalized intersection; G is the effective green time, s; N_l is the number of lanes in a lane group; S_{fr} is the saturation flow rate pc/h/ln; c is the total cycle length (s).

$$S_{fr,CAVs} = s_b \cdot \prod_{i=1}^{n} f_i \tag{2}$$

where: S is the saturation flow rate, veh/h/ln; S_b is the base saturation flow rate, veh/h/ln; f_i is the correction factors.

$$C_{CAVs} = f_{(a)} \cdot a \cdot e^{-f_{(b)} \cdot b \cdot Q_c} \tag{3}$$

where: $C_{e,CAVs}$ is the CAVs' capacity at Roundabout 1 in Fig. 3; a is the intercept parameter equal to 1,350 pc/h for the left entry lane and to 1,380 pc/h for the right entry lane; b is the slope parameter (equal to 0.00092 for the left entry lane and 0.00102 for the right entry lanes; f(a) and f(b) are the corrections for the a and b parameters, respectively [3]. Model parameters were refined based on the alignment of simulated outputs with the ECFs during the calibration process. The green time and traffic light cycle length (see Eq. 1) were collected in the field, while the saturation flow rate was calculated according to [3]. The initial base value for the saturation flow was set at 1,900 vehicles per hour but was subsequently adjusted by factors (f_i) reflecting specific characteristics of the intersection, including lane width, turning movements, heavy vehicles, and CAVs in traffic (see Eq. 2). As CAV-DRs changed, the base saturation flow rate for through movements also varied, while the saturation flow rate for left turns was adjusted with an additional correction factor based on CAV-DRs. Modifications in CAV capacity for right turns were consistent with reference [3]. For example, Fig. 4 shows the ECF surfaces for the left-entry lane (Fig. 4a) at the roundabout, reflecting variations in circulating traffic flow and CAV-DRs.

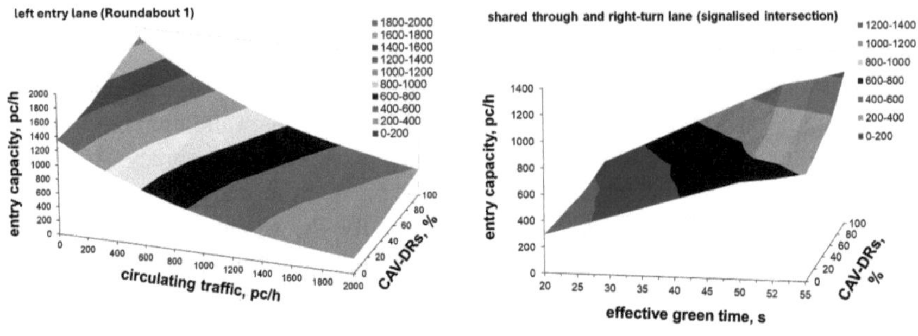

a) (b)

Fig. 4. Surfaces of entry capacity functions for: (a) the left lane at the roundabout; (b) the shared right-turn and through lane group at the signalized intersection.

Additionally, Fig. 4b presents the ECF for the shared right-turn and through lane group at the signalized intersection, demonstrating how effective green time changes in response to different CAV-DRs while keeping the total cycle length constant.

Aimsun version 20 [20] evaluated the impact of various intersection layouts on the performance of CAVs and DoVs. The simulation focused on knee turbo roundabouts, incorporating curbs to facilitate entry lane preselection and minimize lane changes. Traffic data were modelled using geometric features and origin-destination matrices. The total traffic matrix was divided into two distinct parts: one representing a percentage (x) of CAVs and the other representing the remaining percentage of DoVs. Give-way signals were assigned to each entry lane of the roundabout, while a time control plan was developed for the signalized intersection based on field data. Detectors were installed on selected road sections to monitor changes in entry capacity. Ten one-hour trials, involving only conventional vehicles, assessed Aimsun's accuracy in replicating traffic flows, revealing a 13% discrepancy between simulated and field data and indicating the need for model calibration. Assuming high communication reliability, vehicle-to-vehicle communication for CAVs was facilitated by cooperative adaptive cruise control, which enabled the exchange of information to reduce the gap to the vehicle ahead [20]. The Gipps' car-following model analysed vehicle behaviour by evaluating spacing and timing among cars, while the lane-changing model assessed safe gap availability [26]. Gap-acceptance models focus on driver decision-making at intersections based on kinematic parameters. Before calibration, a sensitivity analysis was conducted to understand how changes in model parameters affected simulation outcomes, particularly for global and specific vehicle class parameters. The analysis identified reaction time of moving vehicles as the most sensitive parameter across all models, including reactions at stops and traffic lights. In car-following models, if reaction time is fixed, it synchronizes with simulation steps for all user classes; however, it can also be treated as a variable defined by a probability function tailored to each user class's characteristics. Figure 5 illustrates the sensitivity analysis results, showcasing the variation in entering traffic related to sustainable values for reaction time and reaction time at traffic lights [26].

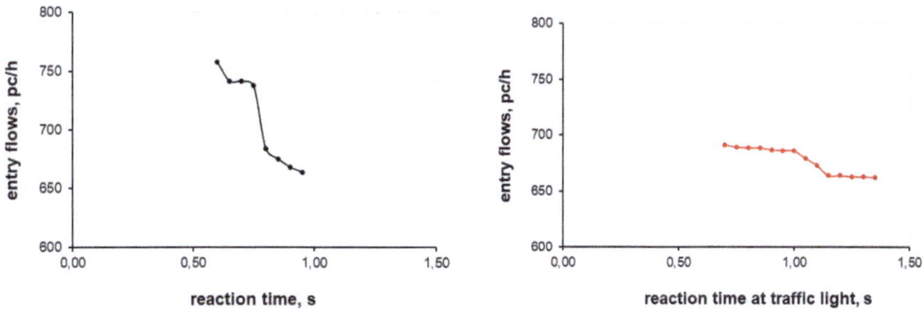

(a) (b)

Fig. 5. Examples of the sensitivity analysis of the model parameters: (a) Reaction time; (b) Reaction time at traffic lights.

The calibration parameters for the baseline case study were:

- The speed limit acceptance, which refers to a driver's adherence to established speed limits, set equal to finely tuned values of 1.30 for DoVs and 1.60 for CAVs in both lane groups, instead of the default value of 1.10.
- The maximum acceleration, measured in m/s^2 and representing the highest acceleration that a vehicle can attain within the network, was set to finely tuned values of 3.20 for DoVs and 3.40 for CAVs in both lane groups, instead of the default value of 3.00.
- The normal deceleration, measured in m/s^2 and used by any vehicle in standard conditions was set to finely tuned values of 3.50 for CAVs in both lane groups, instead of the default value of 4.00, which remains in effect for both lane groups for DoVs.
- The clearance, measured in seconds, refers to the distance between two stationary vehicles; it was set to finely tuned values of 0.50 for CAVs in both lane groups, instead of the default value of 1.00, which remains in effect for both lane groups for DoVs.
- The distance gain, seconds^{-1}, was set to a finely tuned value of 0.60 for CAVs in both lane groups, replacing the default value of 0.45. The time gap leader, seconds, was adjusted to a finely tuned value of 0.50 for CAVs in both lane groups, instead of the default value of 1.50. The time gap follower, seconds, was set to a finely tuned value of 0.50 for all lane instead of the default value of 0.60. The default values were retained for DoVs in both lane groups. These parameters represent the spatial and temporal distances between subsequent vehicles equipped with the CACC module for modelling CAV applications.
- The reaction time, measured in seconds, is a parameter in car-following models that indicates the duration a vehicle takes to respond to changes in speed of the vehicle ahead. It was set to a finely tuned value of 0.85 s for DoVs in the exclusive left-turn lane group, replacing the default value of 0.80 s (and is maintained at 0.80 s in the shared right-turn and through lane group). In CAV modelling, the reaction time is set to 0.62 s in the exclusive left-turn lane group and 0.60 s in the shared right-turn and through lane group.
- The reaction time at a stop, measured in seconds, is a parameter that indicates the duration a subsequent vehicle takes to initiate movement once the preceding vehicle

restarts. It is set to a finely tuned value of 1.75 s for DoVs in the exclusive left-turn lane group, replacing the default value of 1.20 s (while it is set of 1.60 s in the shared right-turn and through lane group). For CAVs, the parameter is set to 1.06 s in the exclusive left-turn lane group and 0.82 s in the shared right-turn and through lane group.

- The reaction time at traffic light, measured in seconds, is a parameter that indicates the duration taken by the first vehicle at the stop line to accelerate in response to a green signal. For DoVs it is set to a finely tuned value of 1.36 s in the exclusive left-turn lane group, replacing the default value of 1.60 s (the value of 1.20 s is set in the shared right-turn and through lane group). For CAVs, the parameter is set to 1.02 s in the exclusive left-turn lane group and 0.73 s in the shared right-turn and through lane group.

Further parameters typically tuned in the context of CAVs operating at intersections included the gap parameter, which measures the distance in seconds between the front bumpers of consecutive vehicles, the safety margin factor, which determines vehicle priorities at intersections, and the sensitivity factor, which relates to the distance between vehicles based on the deceleration assessments of following vehicles. These parameters were maintained at their default values in both lane groups for the baseline case study. For the calibration of parameters described above, the fine-tuned values assumed for the roundabouts are briefly reported as follows:

- The speed limit acceptance for DoVs is set to 0.97 in the left entry lane and 1.00 in the right entry lane, replacing the default value of 1.10, which remains in effect for both entry lanes for CAVs.
- The maximum acceleration, m/s^2, for CAVs is set to 4.00 in both entry lanes, replacing the default value of 3.00, which remains in effect for both entry lanes for DoVs. The normal deceleration, measured in m/s^2, is maintained at the default value of 4.00 m/s^2 for both CAVs and DoVs in both entry lanes.
- The gap in seconds for DoVs is set to 1.33 in the left entry lane and 1.58 in the right entry lane, replacing the default value of 0.00, which remains in effect for both entry lanes for CAVs.
- The safety margin factor for CAVs is set to 0.50 in both entry lanes, replacing the default value of 1.00, which continues to apply for both entry lanes for DoVs.
- The sensitivity factor for CAVs is set to 0.50 in the left entry lane, replacing the default value of 1.00, which continues to apply in the entry lane for CAVs and both entry lanes for DoVs.
- The reaction time, measured in seconds, is set to 0.95 s for DoVs in the left entry lane and 0.86 in the right entry lane, while in CAV modelling, the reaction time is set to 0.67 s in the left entry lane and 0.63 s in the right entry lane, replacing the default value of 0.80 s in all the lanes. The values of the reaction time at stop and traffic light are maintained at their respective default values for both CAVs and DoVs in both entry lanes.

The values of reaction times at the baseline case study and roundabouts were the weighted mean of the corresponding values based on CAV-DRs, with each DR serving as a weight; thus, the fine-tuned values above reported for DoVs were set for 0% of CAVs, while those ones were reported for 100% of CAVs. Further parameters as the clearance, distance gain, time gap leader and time gap follower typically tuned in the context of mixed fleets maintained the default values in both entry lanes for the roundabouts because their adjustment had no significant effect in this case. The most sensitive parameters were calibrated by iteratively adjusting their values to closely align the simulated capacity data with the ECFs, ensuring an accurate representation of traffic behaviour in the model. The GEH index assessed the deviation between simulated data and ECFs, establishing a target for discrepancies under 5 in at least 85% of cases [14]. The GEH index was instrumental in calibrating a two-lane roundabout and resulted in over 90% accuracy for the left lane and 95% for the right lane. The signalized intersection calibration was validated by maintaining a 5% difference between simulated capacity data and ECFs. The percentage difference between these datasets showed the calibration success and validated the calibration procedure.

The calibration results denoted the assertive behaviour of CAVs and DoVs at the baseline case study, in which the values of speed acceptance and maximum acceleration exceeded their respective thresholds. Alternatively, the calibration indicated that DoVs at both entry lanes of the roundabout exhibited somewhat cautious behaviour, with fine-tuned speed acceptance values falling below and reaction time exceeding the thresholds. CAVs exhibited a resolute driving attitude at the roundabout, responding more readily to the leader's actions with higher acceleration than the default values and reaction times and safety margins fell below the thresholds. The tuned CACC-CAV parameters enhanced intra-vehicle communication, enabling these vehicles to drive more resolutely at the signalized intersection than the roundabout, which generally facilitates smoother traffic.

A two-tailed t-test was conducted for the pairs of entry maneuvers included in the same entry mechanism for the baseline case study and roundabout. The t-statistic indicated no significant difference between the means of the two samples (i.e., the simulated data and ECFs), but these deviations were due to the chance. For example, the p-values under the significance level of 0.05 were higher than 0.87 at the baseline case study for CAV-DRs from 20% onwards. Similar results concerned the roundabout. Consequently, the results were consistent with a null hypothesis that is true [27]. The F-statistic confirmed equal sample variances. Additionally, the values of the root mean squared normalized error under 0.04 demonstrated close alignment between the simulated data and target capacities, corroborating the calibration process and the models' effectiveness in replicating the examined phenomenon.

The scattergram analysis was conducted to compare the simulated capacities of the two-lane and the knee-turbo roundabouts and validate the analogy in entry mechanisms where one entering flow is faced by either one or two opposing traffic flows. For example, the scatter plots in Fig. 6 illustrate the relationship between target and simulated capacity, plotted along with the 95% prediction interval [14]. Figure 6a presents capacity data for the left entry lane with a CAV-DR of 60%, while Fig. 6b shows data for the right entry lane with a CAV-DR of 80%. The narrow prediction intervals and high R^2 values indicate a strong positive correlation between the datasets, thereby supporting the extension of the roundabout calibration to the knee-turbo roundabout.

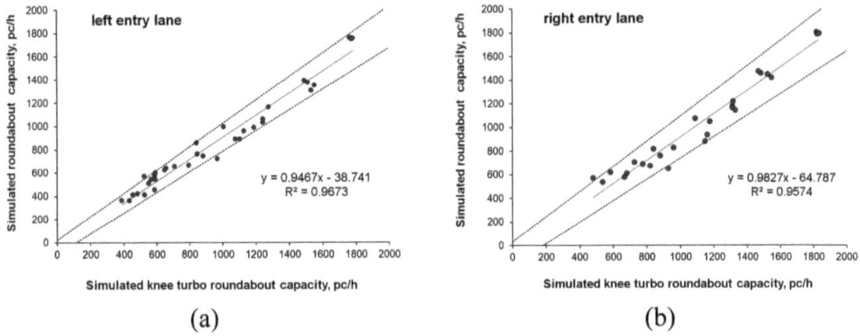

Fig. 6. Scatter plots of the simulated capacities of the two-lane and the knee-turbo roundabouts for: a) the left entry lane with a CAV-DR of 60%; b) the right entry lane with a CAV-DR of 80%.

3 Results

The study utilized a microsimulation-driven approach to evaluate the operational efficiency of CAVs at different intersection designs and traffic conditions. Key performance metrics included entry capacity (vehicles per hour), delay (seconds), and travel times (seconds) across six traffic scenarios with varying CAV-DRs. Entry capacity measured how many vehicles could enter a road section, delay assessed the difference in expected versus actual travel time at intersections, and travel time measured the average duration vehicles spent traveling through the network. Approach 3 in Fig. 3 was the subject entry to show the methodological path. The stacked bar graph in Fig. 7 was used to compare the percentage difference in capacity for each CAV-DR relative to the 100% DoVs scenario for each entry lane of the subject approach considered in the examined intersections. Each segment of the stacked bar in Fig. 7 represents the percentage difference in capacity for each CAV-DR with respect to the 100% DoVs scenario for each entry lane of the subject approach across the baseline study case in Fig. 2a, and Roundabouts 1 and 2 in Fig. 3. Considering the assumptions underlying the study and the uncertainty that can be expected in estimating capacity at roundabouts [28], the simulation results indicated that operational conditions enhanced across all intersections with higher CAV-DRs.

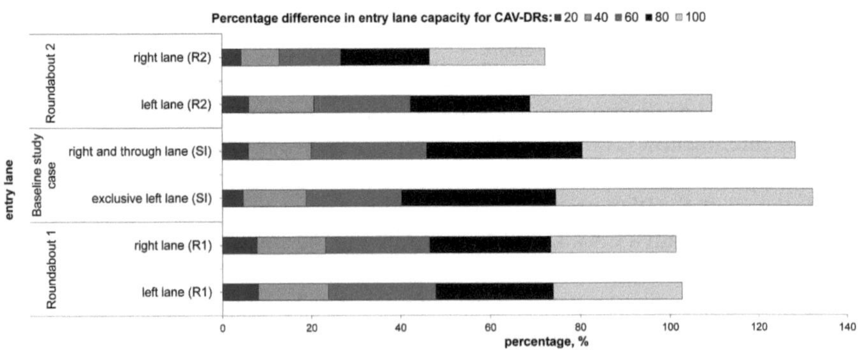

Fig. 7. Percentage difference in capacity for each CAV-DR relative to the 100% DoVs scenario for each entry lane of the subject approach in the examined intersections. Note: R1 stands for roundabout, R2 stands for knee turbo roundabout, SI stands for signalized intersection.

Figure 7 shows greater percentage increases in left entry lane capacity at the roundabout for CAV-DRs, achieving up to 60%. For CAV-DRs at 80%, the signalized intersection's capacity grew by 34%, while roundabouts saw around 27%. With only CAVs, the signalized intersection's capacity exceeded 50% compared to the baseline with only DoVs, and the knee-turbo roundabout experienced approximately 41% growth, with the two-lane roundabout showing a similar trend as observed in the previous case. The roundabout showed greater percentage capacity growth for right entry lanes compared to the baseline study case at CAV-DRs of 20% to 40%. However, at CAV-DRs of 60% to 100%, the baseline study case exhibited a more significant capacity increase than the roundabouts, where operational performance is influenced by traffic matrices [9, 29]. An exclusive left turn lane at the baseline provided higher capacity increments than shared right-turn and through lanes at elevated CAV-DRs. Curbs in the knee-turbo roundabout can enhance capacity for left entry lanes at high CAV-DR values, improving traffic flow, but may also restrict right entry lane capacity by limiting lane spreading on the circulatory roadway. Additionally, traffic lights, by removing the yielding negotiation process, enable higher capacity than the baseline scenario, particularly when fully using CAV-DRs, thereby improving intersection efficiency.

Traffic data collected at the signalized intersection informed the simulation of undersaturated conditions using an unbalanced traffic matrix at the conceptualized intersections. Figure 8 compares the three intersections, highlighting percentage differences in delay measured in seconds for CAV-DRs. The knee turbo roundabout demonstrates a greater percentage decrease in delay for both entry lanes relative to the other intersections. Specifically, the left entry lane shows a reduction up to 65% for 100% CAV-DRs, while the right entry lane ranges up to 46% for 100% CAV-DRs.

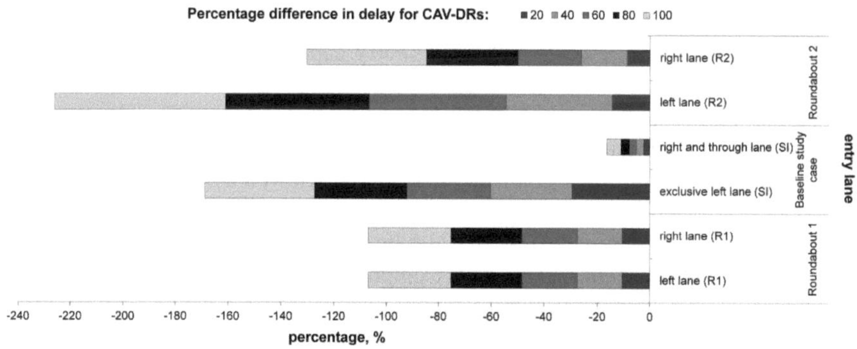

Fig. 8. Percentage difference in delay for each CAV-DR relative to the 100% DoVs scenario for each entry lane of the subject approach in the examined intersections. Note: R1 stands for roundabout, R2 stands for knee turbo roundabout, SI stands for signalized intersection.

Figure 9 illustrates travel time trends for entry lanes, revealing that increased CAV-DRs significantly reduce travel time at the knee turbo roundabout compared to other intersections across various traffic conditions. For a CAV-DR of 20%, the exclusive left turn lane at the signalized intersection showed a 22% reduction in travel time, while the corresponding entry lane at the knee turbo roundabout decreased by only 10%. Notably, for a CAV-DR of 40%, both the left-entry lane at the knee-turbo roundabout and the signalized intersection recorded approximately 23% reductions. The right entry lane at the knee-turbo roundabout exhibited a more substantial reduction, reaching 31% as CAV-DRs increased. Although delay and travel times decreased similarly across the three intersections, the knee-turbo roundabout's geometry positively affected the left entry lane more than the right entry lane, influenced by lane planning and traffic control modes, while lane changes balanced delay reductions in the two-lane roundabout.

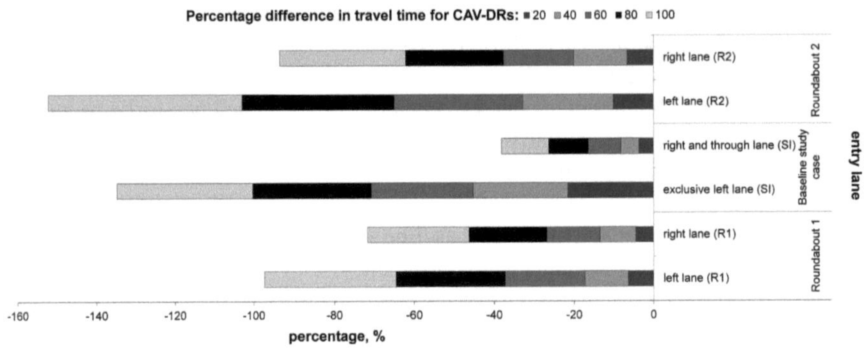

Fig. 9. Percentage difference in travel time for each CAV-DR relative to the 100% DoVs scenario for each entry lane of the subject approach in the examined intersections. Note: R1 stands for roundabout, R2 stands for knee turbo roundabout, SI stands for signalized intersection.

The simulation results showed that intersection schemes significantly affect the performance of connected automated driving, suggesting that certain layouts could enhance smart mobility efficiency as CAVs become more prevalent. CAVs accept smaller traffic gaps than traditional vehicles, enhancing entry capacity and reducing delays with higher deployment rates. Simulations suggested that signalized intersection offers greater capacity improvements than roundabouts, especially at high CAV-DRs. However, two-lane roundabouts facilitate lane changes, which enhances right entry lane capacity. In undersaturated conditions, signalized intersections showed smaller delay reductions compared to roundabouts as CAV numbers increased, particularly in shared lanes. However, the lack of highly automated CAVs in current traffic limits empirical data collection, making outcomes dependent on specific assumptions. The authors emphasize the need for further exploration of diverse intersection cases and control modes to establish a comprehensive performance criterion. This would improve the understanding of contextual factors and help integrate insights related to entry mechanisms into performance evaluations.

4 Conclusions

This paper explores the influence of intersection geometry on the operational performance of connected automated vehicles through a microsimulation-driven approach. It evaluates various intersection designs and traffic conditions relevant to road design, focusing on scenarios involving different CAV deployment rates. Due to a lack of empirical data on high communication-level smart vehicles, assessments rely on assumptions about CAV capabilities. Entry capacity functions were created for calibration purpose. Adjustments to behavioural parameters ensured simulated capacities aligned with target functions for signalized intersections and roundabouts, with statistical analysis confirming the calibration's effectiveness.

The simulation results highlight the operational advantages of fully implementing smart mobility, offering an optimistic outlook for future smart navigation scenarios. Also, the proposed approach facilitates a comparative analysis of alternative intersection designs that incorporate smart vehicle technologies. To enhance findings and model transition scenarios effectively, further case studies should investigate CAV driving behaviour across various road environments and traffic conditions. Although the study's hypotheses may limit the generalizability of its findings, the validation-focused approach has proven effective for evaluating calibration efficacy. Additionally, this research contributes significantly by proposing a novel turbo roundabout design tailored to the traffic and environmental characteristics of the case study context.

The proposed microsimulation-driven framework provides a systematic approach for intelligent mobility management at intersections. The research highlights how smart driving technologies can optimize road usage through effective road design, allowing for practical applications. For instance, adaptive lighting and marking systems—such as dynamic line markings and interactive lights—automatically adjust brightness based on traffic and environmental conditions. Some streetlights activate only when vehicles approach, enhancing visibility. Additionally, advanced solutions aim to optimize interactions among vehicles, pedestrians, and cyclists. Adaptive features or intelligent materials that respond to light and temperature can create dynamic road markings visible at night. These innovations significantly benefit critical locations, particularly at pedestrian crossings and along corridors, ultimately enhancing safety and efficiency within the transportation network. Comparative assessments are essential for cost-benefit analyses of infrastructure projects, aiding in the evaluation of the technical and economic feasibility of proposed designs as we transition sustainably into the smart age. These evaluations provide valuable insights that support informed decision-making during the planning and development stages [1, 30]. In conclusion, the proposed methodology provides transportation planners and road engineers with effective tools to assess intersection design solutions that enhance smart mobility. Successful integration of CAV technologies requires collaboration among research institutions, industry stakeholders, and public engagement within the smart mobility ecosystem, particularly in smart cities. Emphasizing empirical research on CAVs and developing collaborative frameworks for innovative intersection designs will be crucial for promoting sustainable transportation practices and improving overall mobility.

Acknowledgments. The authors acknowledge the support of the Sustainable Mobility Center (Centro Nazionale per la Mobilità Sostenibile — CNMS) under Grant CN00000023 CUP B73C22000760001, and the Project REFERENCES (PNRR MUR M4C2 Initiative 1.3 - Extended Partnerships (EP).

References

1. Pompigna, A., Mauro, R.: Smart roads: a state of the art of highways innovations in the smart age. engineering science and technology. Int. J. **25**, 1–15. paper ID 100986 (2022)
2. Zhang, X., et al.: Vehicle-to-everything communication in intelligent connected vehicles: a survey and taxonomy. Automot. Innov. **8**, 13–45 (2025)
3. National Academies of Sciences: Engineering and Medicine: Highway Capacity Manual 7th Edition: A Guide for Multimodal Mobility Analysis. The National Academies Press, Washington. DC. US (2022)
4. Long, K., et al.: A comprehensive assessment of connected and automated vehicle analytical, modeling, and simulation tools. Transp. Res. Part E: Logist. Transp. Rev. **196**(104007), 1–22 (2025)
5. SAE International. Taxonomy and Definitions for Terms Related to Cooperative Driving Automation for On-Road Motor Vehicles J3216_202107 (2021). https://www.sae.org/standards/content/j3216_202107/
6. Wang, Z., Wu, G., Barth, M.J.: A review on cooperative adaptive cruise control (CACC) systems: architectures, controls, and applications. In: 21st International Conference on Intelligent Transportation Systems (ITSC), Maui, HI, USA, pp. 2884–2891, (2018). https://doi.org/10.1109/ITSC.2018.8569947
7. Peng, J., Shangguan, W., Peng, C., Chai, L.: Uncertainty modeling of connected and automated vehicle penetration rate under mixed traffic environment. Physica A **639**(129640), 1–15 (2024)
8. Richter, T.: Integrated consideration of road sections and intersections. In: Road Planning, Freeways and Country Roads. Springer, Wiesbaden (2023). https://doi.org/10.1007/978-3-658-35189-2_11
9. National Academies of Sciences. Engineering and Medicine: Guide for Roundabouts. The National Academies Press, Washington. DC (2023)
10. Elefteriadou, L.: signalized intersections and arterials. In: An Introduction to Traffic Flow Theory. Springer Optimization and Its Applications, vol. 84, pp. 239–277. Springer, Cham (2024). https://doi.org/10.1007/978-3-031-54030-1_10
11. Liu, P., Fan, W.D.: Exploring the impact of connected and autonomous vehicles on mobility and environment at signalized intersections through vehicle-to-infrastructure (V2I) and infrastructure-to-vehicle (I2V) communications. Transp. Plan. Technol. **44**(2), 129–138 (2021)
12. Tumminello, M.L., Macioszek, E., Granà, A., Giuffrè, T.: Simulation-based analysis of "what-if" scenarios with connected and automated vehicles navigating roundabouts. Sensors **22**(17), 6670, 1–30 (2022)
13. Turbo Roundabouts Informational Primer. Federal Highway Administration. U.S. Department of Transportation. FHWA Safety Program (2022). https://highways.dot.gov/sites/fhwa.dot.gov/files/2022-06/fhwasa20019_0.pdf
14. Barceló, J.: Fundamentals of Traffic Simulation. Springer, London. UK (2010). https://doi.org/10.1007/978-1-4419-6142-6
15. Szarata, A., Ostaszewski, P., Mirzahossein, H.: Simulating the impact of autonomous vehicles (AVs) on intersections traffic conditions using TRANSYT and PTV Vissim. Innov. Infrastruct. Solut. **8**, 164 (2023)
16. Brost, M., et al.: Energy consumption of connected and automated vehicles. In: Elgowainy, A. (ed.) Book Electric, Hybrid, and Fuel Cell Vehicles, Encyclopedia of Sustainability Science and Technology, 2nd edn, pp. 216–224. Springer, New York (2021)
17. Sinha, N., Jain, V., Himanshu, et al.: Synergizing the future: electric vehicles, artificial intelligence, and smart grids. Smart Grids Energy **10**, 17 (2025)

18. Zou, Y., Zheng, F., Liu, C., Liu, X.: Integrated optimization of traffic signals and vehicle trajectories for mixed traffic at signalized intersections: a two-level hierarchical control framework. Transp. Res. Part C: Emerg. Technol. **169**(104884), 1–30 (2024)
19. Gong, X., Lyu, P., Wang, B.: Cooperative motion planning and decision making for CAVs at roundabouts: a data-efficient learning-based iterative optimization method. IEEE Internet Things J. **11**(19), 32205–32220 (2024)
20. Aimsun Next. Version 20 Dynamic Simulator User Manual: TSS-Transport Simulation Systems: Barcelona. Spain (2020)
21. Tumminello, M.L., Zare, N., Macioszek, E., Granà, A.: Smart mobility operation at signalized intersection level in microsimulation environment. In: Macioszek, E., Granà, A., Sierpiński, G. (eds.) TSTP 2024. LNNS, vol. 1315, pp. 153–172. Springer, Cham (2025)
22. Fortuijn, L.G.: Turbo roundabouts: design principles and safety performance. Transp. Res. Rec. **2096**(1), 16–24 (2009)
23. Tollazzi, T.: Alternative Types of roundabouts at development phases. In: Alternative Types of Roundabouts. Springer Tracts on Transportation and Traffic, vol. 6. Springer, Cham (2015)
24. Ministry of Infrastructure: Guidelines for the design of road intersections. Part 3: Roundabouts. WR-D-31-3. Warsaw (2022)
25. Granà, A., Giuffrè, T., Macioszek, E., Acuto, F.: Estimation of passenger car equivalents for two-lane and turbo roundabouts using AIMSUN. Front. Built Environ. **6**(86), 1–16 (2020). https://doi.org/10.3389/fbuil.2020.00086
26. Gipps, P.G.: A behavioural car-following model for computer simulation. Transp. Res. Part B **15**, 105–111 (1981)
27. Dahiru, T.: P - value, a true test of statistical significance? A cautionary note. Ann. Ib. Postgrad. Med. **6**, 21–26 (2008). https://doi.org/10.4314/aipm.v6i1.64038
28. Giuffrè, O., Granà, A., Tumminello, M.L.: Exploring the uncertainty in capacity estimation at roundabouts. Eur. Transp. Res. Rev. **9**(18), 1–13 (2017)
29. Mauro, R., Branco, F.: Comparative analysis of compact multilane roundabouts and turbo-roundabouts. J. Transp. Eng. **136**(4), 316–322 (2010)
30. Schwartz, S.I.: Street Smart: The Rise of Cities and the Fall of Cars. PublicAffairs, New York (2015)

Towards Accurate Vehicle Speed Estimation in Aerial Views: How Localization and Tracking Methods Influence Speed Estimation in UAV-Based Vehicle Monitoring

Muhammad Waqas Ahmed[1(✉)], Muhammad Adnan[1], Muhammad Ahmed[2], Davy Janssens[1], Geert Wets[1], Afzal Ahmed[3], and Wim Ectors[1]

[1] UHasselt, Transportation Research Institute (IMOB), Martelarenlaan 42, 3500 Hasselt, Belgium
{Waqas.ahmed,muhammad.adnan,davy.janssens,geert.wets, wim.ectors}@uhasselt.be
[2] Department of Urban and Infrastructure Engineering, NED University of Engineering and Technology, Karachi 75270, Pakistan
muhammadahmed@neduet.edu.pk
[3] Institute of Transportation Studies, University of Leeds, Leeds LS2 9JT, UK
a.ahmed@leeds.ac.uk

Abstract. Object detection and tracking workflows are increasingly being explored by researchers in transportation sciences because of their practical applications in the field. These algorithms are widely used with unmanned aerial vehicles (UAVs) to develop robust intelligent traffic monitoring systems. However, benchmarking the speed estimates produced by these cutting-edge technologies has not yet been thoroughly conducted. In this study, we present a critical evaluation of various localization methods namely Axis Aligned Bounding Boxes (AABB) and Oriented Bounding Boxes (OBB) of YOLOv8 paired with different tracking models for vehicle speed estimation. Identifying the optimal detector-tracker combination is essential for the advancement of these autonomous traffic monitoring systems. Through this comparative analysis, the fluctuations in speeds estimated by each detector-tracker combination was evaluated on the basis of the Root Mean Squared Error (RMSE), and Signal Noise Ratio (SNR). The experimental results indicate that trackers combined with OBB-enabled detection exhibited lower signal fluctuations than those based on AABB localization. Furthermore, the selection of the tracker significantly impacts the quality of speed estimates.

Keywords: UAVs · Multi-Object Tracking · Traffic Monitoring · Vehicle Speed

1 Introduction

According to a report by the World Health Organization (WHO), there were 1.19 million deaths related to road accidents in 2021. Although this marks a 5% decline since 2010, there is still a lot of work to be done to achieve the goal of zero fatalities from road

E. Macioszek et al. (Eds.): TSTP 2025, LNNS 1769, pp. 106–121, 2026.
https://doi.org/10.1007/978-3-032-14094-4_8

accidents [1]. Additionally, another report on traffic injury prevention estimates that for every 1 km/h increase in speed, the likelihood of crashes rises by 3%, thereby increasing the risk of fatalities from these accidents [2].

The primary challenge in global speed measurement and evaluation is the absence of a scalable and cost-effective solution. While technologies such as inductive loop detectors and magnetic sensors are reliable, they tend to be expensive, making them financially unfeasible for widespread adoption, particularly in low- and middle-income countries (LMICs) [3, 4]. Modern artificial intelligence (AI) based systems hold the promise to enable automation in every sector [5]. In the context of smart cities, AI can greatly enhance various aspects, including traffic enforcement and road safety, ultimately improving citizens' quality of life at a lower cost [6]. However, it is crucial to test and evaluate the efficiency of AI-enabled systems to ensure their effectiveness in real-world scenarios [7]. For example, in cases of violation detection, the AI-based system must be 100% accurate and free from bias to avoid legal complications. Error-prone systems present a significant obstacle to the real-world implementation of these technologies [8, 9].

Unmanned Aerial Vehicles (UAVs) or drones marked a paradigm shift in agriculture, earth observation, meteorology, and even road traffic monitoring [10–13]. UAV technology combined with AI offers an unparalleled aerial advantage over conventional roadside camera systems, offering utility in road asset management, traffic flow analysis, and road safety implementation [14].

Speed estimation of road traffic consists of two steps: pixel calibration or georeferencing, and object detection and tracking. The accuracy of the georeferencing directly influences the error rate; the better the calibration, the lower the error. With the aerial perspective offered by UAVs, the required georeferencing precision is conveniently achievable. Additionally, the accuracy of object localization and tracking plays a crucial role. Variations in bounding boxes during the localization process may cause rapid shifts in the centroid, resulting in inconsistent and noisy speed measurements [15]. These variations can be minimized by exploring different detector–tracker combinations.

This study aims to explore a different combination of detectors and trackers. We utilize footage from UAVs, paired with different object detection and tracking models. Our approach explores traditional axis-aligned bounding box (AABB) localization with state-of-the-art oriented bounding box (OBB) enabled detectors, applied alongside various multi-object tracking algorithms. Each combination produces different results and levels of noise, offering insights into the selection of detectors and trackers for analyzing UAV-based traffic footage.

2 Related Works

With the recent development of state-of-the-art computer vision models, researchers have delved deeply into this domain, particularly with respect to the themes concerning road safety and accident prevention. A practical application of object detection methods for speed estimation was demonstrated by McCraith, Neumann and Vedaldi [16] by mounting a rigid camera on an ego-vehicle and estimated the velocities of other vehicles by estimating the difference in size of their bounding box. Despite achieving an accuracy of 1.28 km/h, the proposed method lacks practicality and does not offer a scalable solution; nonetheless showcases a real-world experiment with AI-based solutions. Similarly, Timofejevs, Potapovs and Gorobetz [17] assessed the potential of cutting-edge object detection algorithms through an experimental setup involving a miniature vehicle operating in a controlled environment. A You Only Look Once (YOLO) object detector is used to detect and estimate the fixed point of the miniature vehicle at each instance. The position at each instance is used to estimate the speed. Meanwhile, an infrared sensor was used to record speeds for validation.

Aerial footage from UAVs coupled with advanced AI algorithms can be advantageous for aerial surveying and traffic monitoring. The advancements in camera technology have enabled the drones to capture high-resolution images at different altitudes [18]. Such footage provides adequate information for the object detection algorithms to work robustly to detect, localize, and track motorized traffic, including cars, vans, buses, and trucks, with high confidence and, under certain conditions, active road users like pedestrians and cyclists as well [19]. Nonetheless, these capabilities resulted in the development of specialized software packages like Datafromsky [20] – a commercial platform for road user detection and localization for traffic analysis and road safety analysis, including estimation of conflict indicators like time to collision (TTC), post encroachment time (PET), and heavy braking. As the technology improves, the increased precision of these methods will eventually enable completely autonomous solutions for road traffic monitoring and violation detection [21].

Autonomous and intelligent traffic monitoring systems can only be successfully implemented if the technology is precise. This precision will soon become a reality with continued research and development in this field. Human feedback plays a crucial role in achieving these goals. A recent study highlights this by presenting an intelligent vehicle violation detection system that incorporates human-computer interaction. Ren [22] employed computer vision technology to detect and register violations for traffic law enforcement and the safety of vulnerable road users. The recorded violations went through a human-computer interaction interface, where the final recorded results are forwarded to the administrator. This enables the communication of useful information and warnings to road users.

The automation in UAV-based traffic systems can be challenging because pixel calibration during an ongoing flight is difficult and can only be done during the post-processing stage. This has previously restricted the development of a flexible solution for calibrating and localizing road traffic footage during the flight. A mitigating solution was presented in our previous research works [23], where the system performed pixel calibration in near real-time using a feature matching algorithm – SIFT [24], and localized and tracked the moving traffic using a Bot-SORT-enabled YOLOv8 algorithm. Although the study showcased a new method for vehicle trajectory localization from stationary and moving drone footage, it did not explore the impact of different detector-tracker combinations on the quality of speed estimates. Similarly, a very recent study by Rathore, Balana, Bhaumik, Pattanaik and Rathore [25] benchmarks different detector-tracker combinations with UAV-based footage. The authors combine EdgeYOLO, YOLOv8s, and Adaptive Sparse Convolutional Networks with Global Context Enhancement (CEASC) for Object Detection on Drone Images with four different sets of trackers to evaluate detection and tracking performance on edge devices. While this study provides valuable insights, its primary focus was on the multi-object tracking accuracy (MOTA) only, rather than the real-world derivatives or byproducts of MOT algorithms. Therefore, this study serves as an extension to these two previous studies by focusing on the speed estimates generated by different detector-tracker combinations. We use footage from UAVs, paired with advanced object detection and tracking methods. Our approach combines traditional axis-aligned bounding box localization with state-of-the-art oriented bounding box methods, applied alongside various multi-object tracking algorithms.

3 Data and Methods

The proposed study aims to assess the impact of various object localization and tracking methods on the accuracy of speed estimates and the level of noise within the data outputs. The speed estimation method begins with calibrating the UAV footage through homography transformation that is achieved by a feature-matching algorithm. Features from the input frame were matched with georeferenced orthomosaics obtained from a previous flight. The experiment footage was collected during daylight hours under clear sky conditions using a DJI Mini Pro 3, at a resolution of 3840x2160 (4k) and a frame rate of 30 frames per second (FPS). During footage collection, the speed of a moving vehicle was recorded using a LiDAR-based speed gun for validation (see Fig. 1).

After pixel calibration, an object detection model – YOLOv8 with two different localization methods was employed: 1) traditional axis-oriented localization and 2) oriented bounding boxes (OBB). These two models were combined with three object trackers: BotSORT, ByteTrack, and BoostTrack++ and the Python script was executed on the T4 GPU of Google Colab environment. The resulting velocities were evaluated for accuracy against point-based LiDAR speed measurements, and the internal noise of the detector-tracker was then plotted for further analysis.

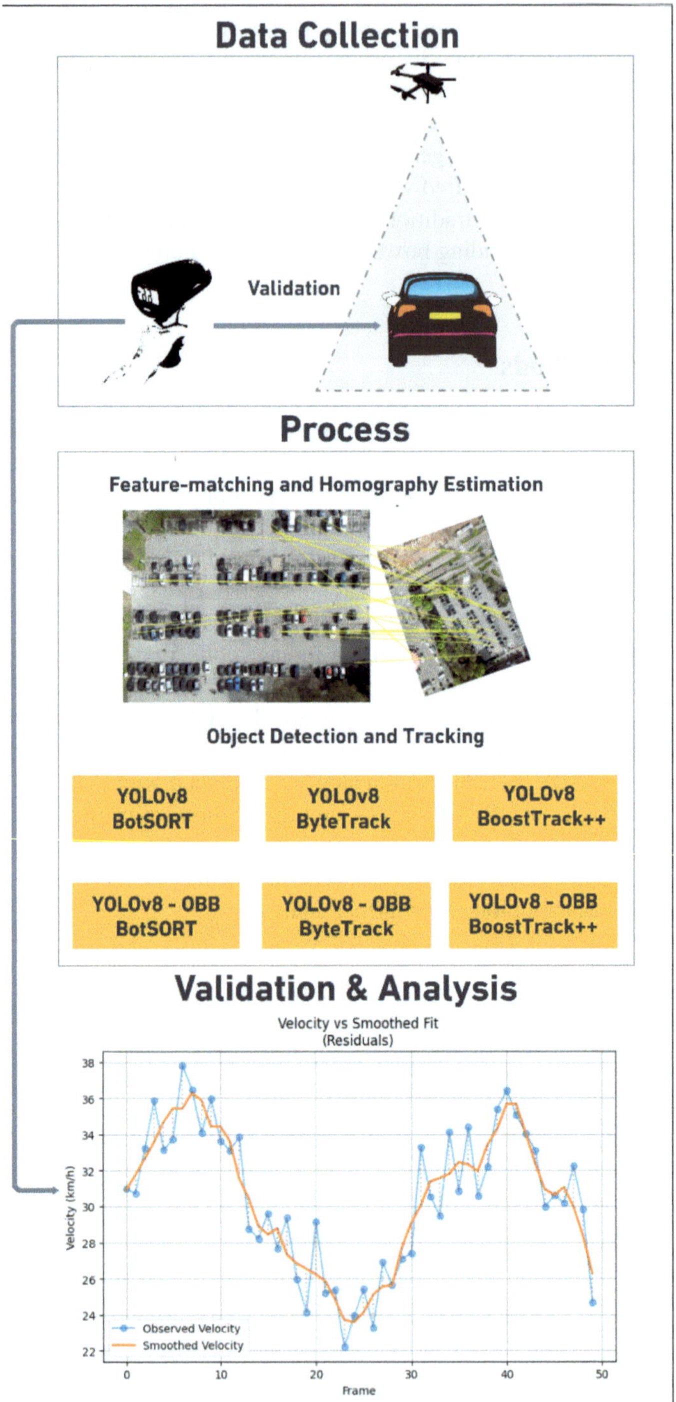

Fig. 1. Illustration of the methodological framework of the study

3.1 Pixel Calibration Using SIFT and Homography Transformation

Pixel calibration is the first step for the speed estimation of a moving road-user. During a drone flight, it can be achieved by utilizing a feature-matching algorithm. In a previous research, Ahmed, Adnan, Ahmed, Janssens, Wets, Ahmed and Ectors [24] were able to calibrate the input frames by calculating homography from a reference orthomosaics using a scale invariant feature transform or SIFT algorithm. Homography transforms points between two planes, which is essential for linking different spatial perspectives. The transformation is based on a 3x3 matrix (Eq. 1), allowing for the conversion of coordinates between these planes.

$$H = \begin{bmatrix} h_{11} & h_{12} & h_{13} \\ h_{21} & h_{22} & h_{23} \\ h_{31} & h_{32} & h_{33} \end{bmatrix} \tag{1}$$

In a homography matrix [h_{11}, h_{12}, h_{21}, h_{22}] describe the affine transformation, whereas, [h_{31}, h_{32}] characterize the perspective transformation, and [h_{13}, h_{23}] represent translation [26]. Traditional methods for homography estimation involve three main steps: feature detection, feature matching, and solving the homography matrix (demonstrated in Fig. 2). Feature extraction algorithms, such as SIFT, SURF, BRIEF, and ORB, are used to identify feature points and generate descriptors represented as vectors. Similarity measures like Hamming or Euclidean distance help match these feature points, while robust estimation techniques like random sample consensus (RANSAC) enhance the reliability of the homography matrix estimation [27]. SIFT has immunity to seasonality due to its ability to match features despite variations in pixel intensities, making it a reliable choice for feature-matching between images taken at different temporal periods [28].

Fig. 2. Feature-matching and homography initialization using SIFT.

SIFT was developed by Lowe [29] in 2004, and was designed to detect distinctive invariant features from images, allowing for reliable matching between two views of the same image. The SIFT process can be divided into four stages. The first step is scale-space extrema detection, where a Gaussian scale-space pyramid is created. This involves generating successive blurred images through convolution with Gaussian functions, thus forming multiple octaves. The parameter sigma establishes the initial Gaussian blur for the first level of each octave. In this stage, the Difference of Gaussian (DoG) method is applied, calculating the difference between images at different scales within an octave. The parameter n-scale indicates the number of scales sampled in each octave. While increasing the number of scales can result in more detected key points, it may also decrease their stability. A balance must be struck between the total number of key points and the stable key points found. The second step is feature localization, which involves a detailed fitting process for each extremum to determine its location, scale, and the ratio of principal curvatures. Keypoints identified in areas of low contrast or along edges are filtered out to ensure that the feature set remains robust. Next is the orientation assignment phase, where an orientation is assigned to each key point based on the local image gradients. This step is crucial for achieving rotational invariance in the final feature descriptors. Finally, during the feature descriptor creation stage, SIFT computes a descriptor for each keypoint invariant to residual variations. Each descriptor is a vector with 128 elements encapsulating pixel properties within its neighborhood [30].

For calibration of the footage, a preconstructed reference image with universal traverse Mercator (UTM) projection system was used. This reference orthomosaic was created from a previous flight using a DJI mini 3 Pro drone at an altitude of 90 m, with a 48-megapixel camera at a top-down angle. The imagery was processed using a web-based application called WebODM, a product of OpenDroneMap. Feature-matching was done using SIFT, matching features between a template image and the input frame, and the projection of the reference image was extracted onto the input footage by the matched keypoints and descriptors (as shown in Fig. 2). For the quality assurance of descriptors, Lowe's ratio test is enforced, and only the descriptors passing the criteria are used for homography estimation [31].

3.2 Object Detection and Localization

Object Detection and Axis-Aligned Bounding Boxes (AABB)

The YOLO algorithm represents a significant leap forward in object detection, introducing a single-stage detection. YOLO divides the input image into a grid, simultaneously predicting bounding boxes and class probabilities. This grid-based approach allows for faster inference, enabling near-real-time and real-time applications [32]. The YOLO algorithm has seen numerous enhancements over the years, resulting in several versions, each offering improvements in detection precision and inference speed. The eighth generation of YOLO architecture has drawn considerable attention for its superior performance in identification tasks compared to its predecessors and even successors in certain use cases [33]. The researchers have rigorously evaluated YOLOv8 in diverse experimental setups. The rationale for using YOLOv8 instead of its successors was its superior performance for detecting larger road user classes, such as cars, vans, and trucks, which are our classes of interest in this study, as the speed gun measurements were recorded for a specific vehicle. The 10th-generation YOLO architecture exhibits better detection capabilities for smaller objects such as pedestrians [33].

The same version of YOLO architecture (version 8) was used for AABB and OBB to enable a fair comparison. Using different models could produce different results; therefore, the detector-tracker combination would not have been comparable. The AABB weights were trained on the VisDrone2019-DET dataset, designed explicitly for detecting low-lying objects like road traffic [34]. The AP @0.5 of 64% was recorded for the class of interest: cars, which is adequate, as all the vehicles in the experiment footage remain detected and localized throughout the sample footage, and consistency was manually validated. Different combinations of AABB model weights and object trackers were created, including the default Bot-SORT, ByteTrack, and Boostrack++. The trained model was applied to the calibrated aerial footage, and the positions of vehicles were detected and localized across all frames. This facilitates the velocity estimation.

Oriented Bounding Boxes (OBB)

Recent YOLO architectures offer a new feature called oriented bounding boxes (OBB), capturing the orientation of detected objects. Earlier YOLO versions localized objects using AABB, a method incapable of capturing the true orientation of vehicles. This often resulted in fluctuations in detection, causing centroidal shifts. The OBB models are trained on the DOTA dataset, which incorporates an additional parameter - the object's angle (as shown in Fig. 3) - to locate objects precisely [35].

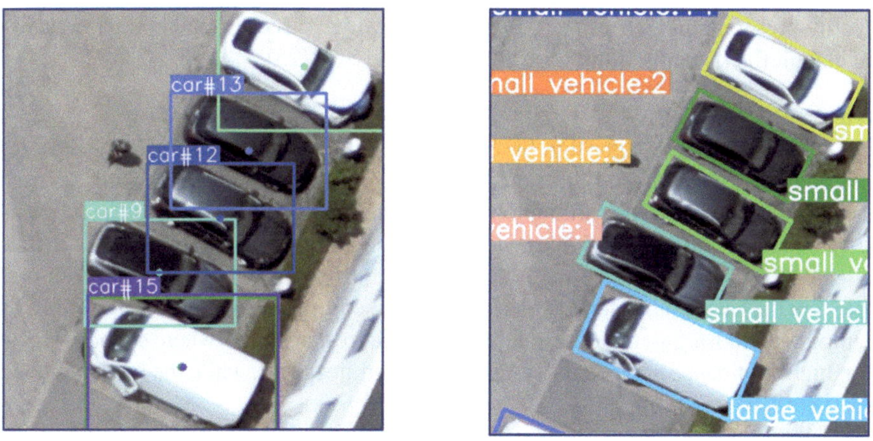

(a) (b)

Fig. 3. The figures illustrate (a) Axis Aligned Bounding Boxes (AABB) and (b) Oriented Bounding Boxes (OBB).

The AABB-based inference method struggled to efficiently localize elongated and rotated objects. In contrast, OBB-enabled inference provides more robust localization by reducing overlap and improving isolation of objects in crowded environments. Additionally, OBBs also stabilize the confidence scores of detections, thereby minimizing fluctuations in bounding box sizes that can cause shifts in the object's centroid across input frames [36]. Previously, these centroidal shifts resulted in rapid movements, leading to inaccuracies in speed measurements [15]. The localization format of OBB is in xywhr format, where x and y are center coordinates, w and h are the width and height of the bounding boxes, and r is the rotation angle (θ) [37].

3.3 Multi-object Tracking (MOT) and Velocity Estimation

Object tracking is a process after object detection, where each detected object is assigned a unique identifier and tracked across frames. Multi-object tracking (MOT) involves tracking multiple objects within a video frame and representing them as trajectories. Over the years, researchers have developed several state-of-the-art MOT algorithms and consistently evaluated their performance. However, challenges such as occlusion, appearance variability, and camera motion often result in tracking inaccuracies. Additionally, there is a noticeable lack of literature with critical appraisals of these models across different video platforms and real-world scenarios [38, 39]. For this experiment, the MOT algorithms integrated within the Ultralytics platform, specifically BotSORT and ByteTrack, were utilized [25], along with an external MOT algorithm called BoostTrack + + [40].

Bot-SORT

BotSORT is a robust state-of-the-art MOT algorithm presented by Aharon, Orfaig and Bobrovsky [41] in 2022. The preceding MOT algorithms, like SORT and DeepSORT, showed limitations, particularly in terms of bounding box accuracy and camera motion. Bot-SORT addresses these limitations by incorporating camera motion compensation techniques for improved localization. Additionally, BoT-SORT also introduced reidentification (ReID) methods for object association, improving the detection accuracy and identity preservation metrics [42]. Bot-SORT is widely used for object tracking and is the default tracking algorithm for YOLO by Ultralytics [25].

ByteTrack

ByteTrack is an MOT algorithm based on the tracking-by-detection (TBD) model. It uses bounding boxes from object detection and performs trajectory tracking. In traditional TBD algorithms, discarding low-confidence detections is a standard practice. To overcome low-scoring detections, ByteTrack presented a novel data association method called BYTE. This approach establishes trajectories from high-confidence detections and filters out the background noise by discarding low-scoring detection boxes. Basically, the method uses both high-confidence and low-confidence detections, where bounding boxes with high confidence are meant to establish target trajectories, while the low-confidence boxes are removed [43].

Boostrack++

MOT relies on the accurate selection of true-positive detected bounding boxes. This challenge is primarily addressed by implementing a two-stage association process and utilizing low-confidence detections in the second stage. Ideally, an MOT algorithm should prioritize true positive detections to minimize identity switches. Boosttrack++ enhances the similarity measure by incorporating a detection-tracklet confidence score, which implicitly favors pairs with high detection and tracklet confidence during one-stage association. To reduce ambiguity associated with using intersection over union (IoU), Mahalanobis distance and shape similarity are combined to improve the overall similarity measure. As this method is integrated with appearance similarity, it surpasses all standard benchmark solutions on the MOT17 and MOT20 datasets [40]. In this experiment, Boostrack++ was used with pretrained ReID weights, OSNet trained on the MSMT17 dataset [44].

3.4 Speed Estimation

After the object detection and trajectory estimation phase from the georeferenced UAV footage, the vehicle positions with respect to time are estimated. To estimate the instantaneous velocity of a moving object, we analyzed its positional data in terms of Northing and Easting coordinates with the 'meter' as unit, recorded at discrete time intervals. The velocity vector at each frame was computed as the first-order finite difference of position with respect to time [45]. The method enables an approximation of continuous motion in two-dimensional space using discrete spatial-temporal data (see Eq. 2).

$$S = \sqrt{\left(\frac{E_i - E_{i-1}}{\Delta t}\right)^2 + \left(\frac{N_i - N_{i-1}}{\Delta t}\right)^2} \quad (2)$$

where:

$N_i and E_i$ are the northing and eastings of a track at frame i and $\Delta t = t_i - t_{i-1}$.

3.5 Error Estimation and Signal-to-Noise Ratio

For error and signal noise estimation, a baseline was established using the Savitzky–Golay filter. The Savitzky–Golay smoothing and differentiation filter optimally fits a set of data points to a second-order polynomial or quadratic line as a convoluted function [46]. The resulting fitted line captures the trend for each track while reducing the high-frequency noise (illustrated by Fig. 4). The residuals are calculated to identify noise or deviation from the underlying trend by subtracting the smoothed signal from the raw velocity values (see Eq. 3).

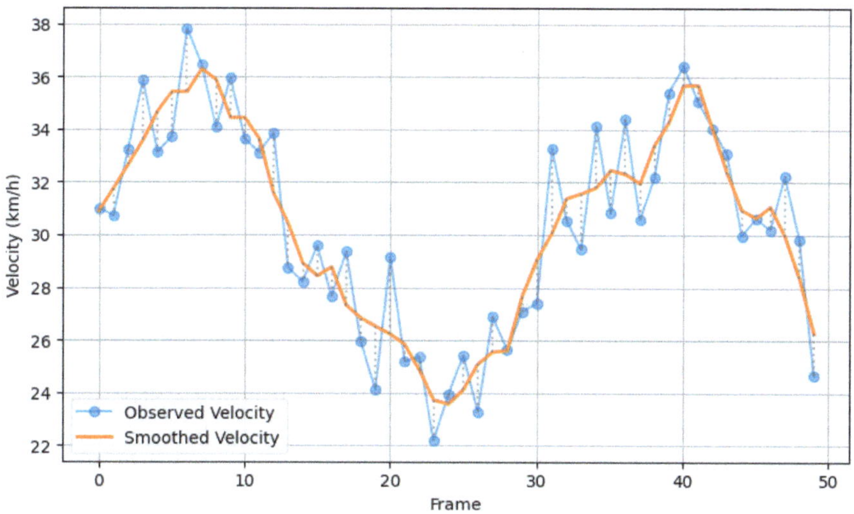

Fig. 4. An example illustration of Savitzky–Golay filter on vehicle speeds for residual estimation.

A Signal Noise Ratio (SNR) and Root Mean Square Error (RMSE) are commonly used signal evaluation indices [47]. SNR is defined as the ratio between signal and noise power. The larger the SNR, the better the data smoothing. SNR [48] is defined as Eq. (6). In this experiment, the time series of raw speeds (km/h) and smoothed speeds after the Savitzky–Golay filter are plotted, and residual noise is calculated by residuals r_i (Eq. 3).

$$r_i = v_i - s_i \tag{3}$$

where v_i is raw velocity and s_i is smoothed velocity, the signal power P_s is calculated by Eq. (4):

$$P_s = \frac{1}{N}\sum_{i=1}^{N} {s_i}^2 \tag{4}$$

Similarly, the noise power is estimated as Eq. (5)

$$P_n = \frac{1}{N}\sum_{i=1}^{N} {r_i}^2 \tag{5}$$

Finally, the signal to noise ratio is described as

$$SNR = 10 \times log_{10}\left(\frac{P_s}{P_n}\right) \tag{6}$$

4 Results and Discussion

In the experimental footage, 20 vehicles were detected, localized, and tracked for 300 consecutive frames. The localized trajectories were used to estimate the positional change or speed. Reference reading from the speed gun was used to check whether all of the detector-tracker combinations estimated the speeds correctly. In the speed plots (illustrated in Fig. 5), track 2 hovers around the reference measurements, i.e., 26km/h. This confirms that all of the models are able to capture speed accurately.

From further analysis of other tracks (parked vehicles), a lot of signal noise is generated. A high rate of fluctuations is observed in AABB object detectors, as the figure shows a parked vehicle registering a significant amount of noise. It was also observed that the default parameters of BotSORT and ByteTrack are not as robust as Boosttrack++, where external reidentification weights significantly enhance the tracking, as track 20 was tracked from the 47th frame onwards with the BotSORT and ByteTrack trackers.

The results were further evaluated for all six detector-tracker combinations on the basis of RMSE and SNR. RMSE was used as a primary metric to assess and compare the absolute performance of a detector-tracker pair. In contrast, SNR was used to understand which method provides the cleanest signal relative to its own scale. The AABB-based YOLO trackers exhibited relatively high average speeds even with parked vehicles, because their bounding boxes must continually resize to contain the vehicle's changing orientation and perspective. Every time the car turns, pitches, or is angularly parked, the box expands or contracts, shifting its centroid even when the true translation is modest. Boosttrack's self-adapting reidentification mechanism reduces ID switches, jitters, and drift, yielding the lowest RMSE (1.28 km/h) and the highest SNR (9.37 dB)

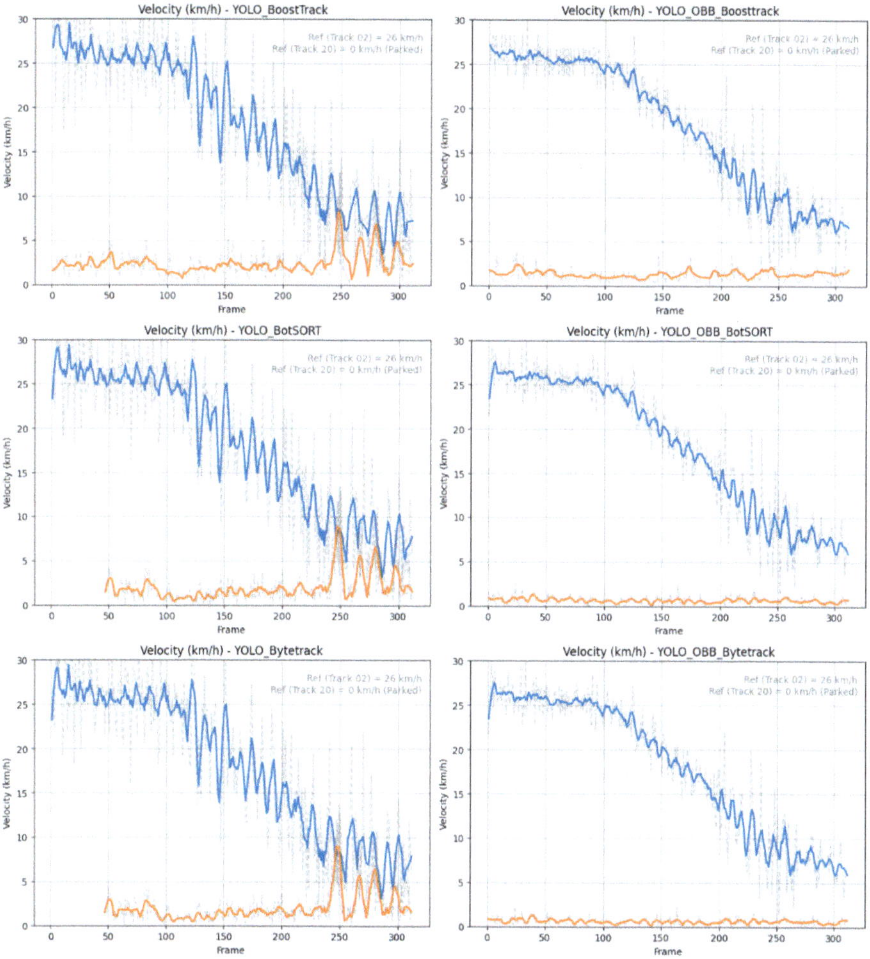

Fig. 5. Illustrates the speed estimates (in km/h) of two tracks (a moving and a parked car) (smoothened with Sav-Gol) for all six detector – tracker combinations [Left: AABB, Right: OBB]. The single-point reference speed measured by the speed gun was 26 km/h.

among the AABB methods (see Table 1). In contrast, BotSORT and Bytetrack, with their more aggressive motion predictions and association logic, tend to overshoot more often, resulting in larger frame-to-frame displacements (RMSE ~2.0 km/h) and an SNR in the mid-8 dB.

Switching to oriented (OBB) bounding boxes dramatically cleans up both speed estimates and noise. Because the OBB rotates with the vehicle, centroid movement more closely tracks true translation. As a result, all three OBB variants report much lower average speeds as the parked vehicles are not registering movements. The RMSE is significantly lower (0.31–0.61 km/h) compared to their AABB counterparts. Among all three OBB algorithms, YOLO-OBB Bytetrack delivers the ideal performance with a low RMSE and high SNR. Boostrack++ demonstrated higher RMSE compared to its

Table 1. Summary of each algorithmic combination's performance.

Algorithm	RMSE (km/h)	SNR (dB)
YOLO Boosttrack++	1.28	9.37
YOLO BotSORT	2.03	8.34
YOLO Bytetrack	2.00	8.28
YOLO-OBB Boosttrack++	0.61	9.90
YOLO-OBB BotSORT	0.31	8.73
YOLO-OBB Bytetrack	0.32	9.54

counterparts; however, its performance can be improved by experimentation with the hyperparameters and reidentification weights. In these experiments, the same settings for both AABB and OBB models were used to ensure a fair comparison. To sum up, the OBB-enabled algorithms can greatly boost the vehicle detection, tracking, and speed estimation with cleaner trackers and minimal noise.

As the technology advances, object detection and detection algorithms are becoming increasingly sophisticated and better suited to assist road traffic monitoring systems. This progress could ultimately lead to the development of a fully automated, low-cost solution for identifying instances of over-speeding, violation detection, and making the vision of zero traffic fatalities a reality.

5 Conclusion

In this paper, we experimented with different combinations of detection and tracking models to estimate vehicle speeds for road traffic monitoring and safety applications. Specifically, the experiment tested two localization methods—Axis-Aligned Bounding Box (AABB) and Oriented Bounding Box (OBB)—paired with three tracking algorithms. The results confirmed that any tracker using the OBB-based detector outperformed those using AABB, producing better results with less signal noise.

While the study focuses on accuracy and noise, it is worth noting that running these algorithms is nearly identical in terms of computational effort; none of the tested combinations demonstrated a significant difference in processing speed. This highlights that accuracy improvements can be prioritized without sacrificing efficiency. The study confirms that continued exploration of technology is essential for identifying models that are more suitable for this application, ultimately leading to fully automated road traffic monitoring and violation detection systems. These investigations are crucial because noise in the speed estimates can result in false positives when detecting instances of speeding. Improving these smart solutions could be beneficial for LMICs, as they provide a cost-effective alternative to expensive equipment that requires regular maintenance. AI technology offers a more affordable option with the potential for global implementation.

Acknowledgements. The authors would like to express their sincere gratitude to the BOF/BILA program at UHasselt for co-funding this research under grant # 14406 (BOF24BL02).

References

1. World Health Organization: Global status report on road safety 2023: summary. World Health Organization (2023)
2. Peden, M.M.: World report on road traffic injury prevention. World Health Organization (2004)
3. Gheorghiu, R.A., Iordache, V., Stan, V.A.: Urban traffic detectors–comparison between inductive loop and magnetic sensors. In: 2021 13th International Conference on Electronics, Computers and Artificial Intelligence (ECAI), pp. 1–4. IEEE (2021)
4. Kostakos, V., Ojala, T., Juntunen, T.: Traffic in the smart city: exploring city-wide sensing for traffic control center augmentation. IEEE Internet Comput. **17**, 22–29 (2013)
5. Shekhar, S.S.: Artificial intelligence in automation. Artif. Intell. **3085**, 14–17 (2019)
6. Elloumi, M., Dhaou, R., Escrig, B., Idoudi, H., Saidane, L.A.: Monitoring road traffic with a UAV-based system. In: 2018 IEEE Wireless Communications and Networking Conference (WCNC), pp. 1–6. IEEE (2018)
7. Deswal, S., Srivastava, S., Gulati, N.: AI based real time vehicle speed detection using deep learning. In: Advances in Networks, Intelligence and Computing, pp. 82–90. CRC Press (2024)
8. Franklin, R.J.: Traffic signal violation detection using artificial intelligence and deep learning. In: 2020 5th International Conference on Communication and Electronics Systems (ICCES), pp. 839–844. IEEE (2020)
9. Ravish, R., Rangaswamy, S., Char, K.: Intelligent traffic violation detection. In: 2021 2nd Global Conference for Advancement in Technology (GCAT), pp. 1–7. IEEE (2021)
10. Tewes, A.: Investigating the Potential of UAV-Based Low-Cost Camera Imagery for Measuring Biophysical Variables in Maize. Universitäts-und Landesbibliothek Bonn (2018)
11. Cox, T.H.: Earth observations and the role of UAVs: a capabilities assessment (2006)
12. Hervo, M., Romanens, G., Martucci, G., Weusthoff, T., Haefele, A.: Evaluation of an automatic meteorological drone based on a 6-month measurement Campaign. Atmosphere **14**, 1382 (2023)
13. Outay, F., Mengash, H.A., Adnan, M.: Applications of unmanned aerial vehicle (UAV) in road safety, traffic and highway infrastructure management: recent advances and challenges. Transp. Res. Part A: Policy Pract. **141**, 116–129 (2020)
14. Chen, C., et al.: YOLO-based UAV technology: a review of the research and its applications. Drones **7**, 190 (2023)
15. Lin, J.-P., Sun, M.-T.: A YOLO-based traffic counting system. In: 2018 Conference on Technologies and Applications of Artificial Intelligence (TAAI), pp. 82–85. IEEE (2018)
16. McCraith, R., Neumann, L., Vedaldi, A.: Real time monocular vehicle velocity estimation using synthetic data. In: 2021 IEEE Intelligent Vehicles Symposium (IV), pp. 1406–1412. IEEE (2021)
17. Timofejevs, J., Potapovs, A., Gorobetz, M.: Algorithms for computer vision based vehicle speed estimation sensor. In: 2022 IEEE 63th International Scientific Conference on Power and Electrical Engineering of Riga Technical University (RTUCON), pp. 1–6 (2022)
18. Ryan, J., Maloney, D., Quinn, B.: What Does a Drone See?: How Aerial Data Resolution Impacts Data Protection (2019)
19. Yu, H., et al.: The unmanned aerial vehicle benchmark: object detection, tracking and baseline. Int. J. Comput. Vision **128**, 1141–1159 (2020)
20. Adamec, V., Herman, D., Schullerova, B., Urbanek, M.: Modelling of traffic load by the DataFromSky system in the smart city concept. In: Lopes, N.V.M. (ed.) Smart Governance for Cities: Perspectives and Experiences, pp. 135–152. Springer, Cham (2020)

21. Dilek, E., Dener, M.: Computer vision applications in intelligent transportation systems: a survey. Sensors **23**, 2938 (2023)
22. Ren, Y.: Intelligent vehicle violation detection system under human-computer interaction and computer vision. Int. J. Comput. Intell. Syst. **17**, 40 (2024)
23. Ahmed, M.W., et al.: From stationary to nonstationary UAVs: deep-learning-based method for vehicle speed estimation. Algorithms **17**, 558 (2024)
24. Ahmed, M.W., et al.: Near real-time privacy protection: automated location-dependent video blurring in UAV live-streams. Transp. Res. Procedia **84**, 201–208 (2025)
25. Rathore, P.S., Balana, A., Bhaumik, R., Pattanaik, V., Rathore, P.: Benchmarking object detection and tracking for UAVs: an algorithmic comparison. In: 2024 IEEE International Conference on Vehicular Electronics and Safety (ICVES), pp. 1–6. IEEE (2024)
26. Guan, B., Vasseur, P., Demonceaux, C., Fraundorfer, F.: Visual odometry using a homography formulation with decoupled rotation and translation estimation using minimal solutions. In: 2018 IEEE International Conference on Robotics and Automation (ICRA), pp. 2320–2327. IEEE (2018)
27. Luo, Y., et al.: A review of homography estimation: advances and challenges. Electronics **12**, 4977 (2023)
28. Van Quan, N., Luyen, D.D., Anh, P.H., Thu, P.T.H., Thanh, N.C., Thai, V.D.: Speeding up SIFT for faster feature-based UAV visual localization. In: International Conference on Advances in Information and Communication Technology, pp. 83–90. Springer, Cham (2023)
29. Lowe, G.: Sift-the scale invariant feature transform. Int. J. **2**, 2 (2004)
30. Castillo-Carrión, S., Guerrero-Ginel, J.-E.: SIFT optimization and automation for matching images from multiple temporal sources. Int. J. Appl. Earth Obs. Geoinf. **57**, 113–122 (2017)
31. Kaplan, A., Avraham, T., Lindenbaum, M.: Interpreting the ratio criterion for matching SIFT descriptors. In: Computer Vision–ECCV 2016: 14th European Conference, Amsterdam, The Netherlands, October 11–14 2016, Proceedings, Part V 14, pp. 697–712. Springer, Cham (2016)
32. Boudjit, K., Ramzan, N.: Human detection based on deep learning YOLO-v2 for real-time UAV applications. J. Exp. Theor. Artif. Intell. **34**, 527–544 (2022)
33. Sundaresan Geetha, A., Alif, M.A.R., Hussain, M., Allen, P.: Comparative analysis of YOLOv8 and YOLOv10 in vehicle detection: performance metrics and model efficacy. Vehicles **6**, 1364–1382 (2024)
34. Chen, G., et al.: VisDrone-MOT2021: the vision meets drone multiple object tracking challenge results. In: Proceedings of the IEEE/CVF International Conference on Computer Vision, pp. 2839–2846 (2021)
35. Feng, S., Huang, Y., Zhang, N.: An Improved YOLOv8 OBB model for ship detection through stable diffusion data augmentation. Sensors **24**, 5850 (2024)
36. Zand, M., Etemad, A., Greenspan, M.: Oriented bounding boxes for small and freely rotated objects. IEEE Trans. Geosci. Remote Sens. **60**, 1–15 (2021)
37. Kim, K.H., Koo, H.-W., Lee, B.-J.: Deep learning-based localization and orientation estimation of pedicle screws in spinal fusion surgery. Korean J. Neurotrauma **20**, 90 (2024)
38. Hassan, S., Mujtaba, G., Rajput, A., Fatima, N.: Multi-object tracking: a systematic literature review. Multimedia Tools Appl. **83**, 43439–43492 (2024)
39. Amosa, T.I., et al.: Multi-camera multi-object tracking: a review of current trends and future advances. Neurocomputing **552**, 126558 (2023)
40. Stanojević, V., Todorović, B.: BoostTrack++: using tracklet information to detect more objects in multiple object tracking. arXiv preprint arXiv:2408.13003 (2024)
41. Aharon, N., Orfaig, R., Bobrovsky, B.-Z.: Bot-sort: robust associations multi-pedestrian tracking. arXiv preprint arXiv:2206.14651 (2022)

42. Pereira, R., Carvalho, G., Garrote, L., Nunes, U.J.: Sort and deep-SORT based multi-object tracking for mobile robotics: evaluation with new data association metrics. Appl. Sci. **12**, 1319 (2022)
43. Sheng, B., Sun, Q., Sun, Y., Hua, Z., Zhang, Y., Tao, J.: A study on improved YOLOv7-pose and bytetrack for dual person tracking. In: 2023 29th International Conference on Mechatronics and Machine Vision in Practice (M2VIP), pp. 1–6 (2023)
44. Zhou, K., Xiang, T.: Torchreid: a library for deep learning person re-identification in PyTorch. arXiv preprint arXiv:1910.10093 (2019)
45. Porter, M., Whitton, M., Kriellaars, D.: Assessing driving with the global positioning system: effect of differential correction. Transp. Res. Rec. **1899**, 19–26 (2004)
46. Luo, J., Ying, K., Bai, J.: Savitzky-Golay smoothing and differentiation filter for even number data. Sig. Process. **85**, 1429–1434 (2005)
47. Zhao, N., et al.: An integrated method for tunnel health monitoring data analysis and early warning: Savitzky-Golay smoothing and wavelet transform denoising processing. Sensors **23**, 7460 (2023)
48. Van Biesen, J.: Measurements of Signal-to-Noise Ratio (SNR) of industrial x-ray film. In: Boogaard, J., van Dijk, G.M. (eds.) Non-destructive Testing, pp. 117–119. Elsevier, Oxford (1989)

A Virtual Approach to Evaluate Vehicular Flows to Promote Sustainable and Safe Mobility in Small and Smart Communities

Giuseppe Cappelli[1(✉)], Sofia Nardoianni[1], Mauro D'Apuzzo[1], and Vittorio Nicolosi[2]

[1] University of Cassino and Southern Lazio, Via G. Di Biasio 43, 03043 Cassino, Italy
{giuseppe.cappelli1,sofia.nardoianni,dapuzzo}@unicas.it
[2] University of Rome "Tor Vergata" , Via del Politecnico, 1, 00133 Rome, Italy
nicolosi@uniroma2.it

Abstract. A key feature of Smart and Sustainable Mobility regards Road Safety Improvements. One of the big challenges is to make small communities safer and smarter, where the main issue refers to the lack of data availability about mobility and traffic demand. Without these data, applying some sustainable, smart, and safe mobility solutions is quite difficult. To overcome this limitation, in this paper, a methodology to forecast vehicular traffic flow is proposed. The proposed method enables users to evaluate vehicular flows by combining travel times retrieved from a navigation software with the Greenshields, BPR, and free flow speed models. The main advantage of the model is to provide a simple tool to be used for virtual collection of traffic data to be used for the calibration and validation of travel-demand macroscopic and/or microscopic models. The method has been tested in the city of Cassino, providing an agreement with real traffic flows with a satisfactory Pearson coefficient of 0.9216). It is expected that this method could be an important tool in the hands of city managers and professionals in making smarter and safer communities.

Keywords: Smart Mobility · Sustainable Mobility · Road Safety

1 Introduction

The urban configuration of cities worldwide has changed in recent years. In 1999, the world population reached 6 billion, and greater growth is forecast for 2040, when the number will reach 9 billion people [1]. Although the quality of life in urban areas is considered to be better than in rural areas, the continued demand for resources, new infrastructure, and energy, which are bounded by unsustainable global growth, has and will effectively weaken the ecosystems and biodiversity of many areas, making immediate action necessary to make cities more sustainable, and smarter. The greater demand for the transportation sector, private, public, and freight-related, that will face this growth, will pose questions on creating and optimizing algorithms for vehicles and people, and traffic management to reduce congestion [2]. Smart mobility technology and solutions could improve traffic congestion, reduce travel times, and save fuel; give users new and

E. Macioszek et al. (Eds.): TSTP 2025, LNNS 1769, pp. 122–137, 2026.
https://doi.org/10.1007/978-3-032-14094-4_9

alternative mobility solutions and guarantee faster travel, to reduce costs for the environment and users; on the other hand, for administrators, smart mobility could help in better traffic management with real-time monitoring and decisions, and improve route planning [3].

As suggested by Chouraby et al. [4], smart mobility is a crucial and fundamental feature of smart cities, and the key attributes are flexibility, efficiency, integration, sustainability, security and safety, social benefits, automation, connectivity, accessibility, and user experience [2]. But what is the definition of a smart city? According to Lai et al. [5], there is no single definition, and the reason is due to several identities that are involved in the process, but also the several different technologies that are implemented in the cities [6]. In addition to that, adjectives such as "smart" and "sustainable" seem to be to a certain degree related and exchangeable: smart and sustainable cities urgently need new adequate, technological, and digital infrastructures [7] to give road users more accessible, direct, and easier mobility solutions and services [8]. Despite this uncertain definition of a smart city, the European Commission tries to explain: "*A smart city is a place where traditional networks and services are made more efficient with the use of digital solutions for the benefit of its inhabitants and business*" [9]. As suggested by Benevolo et al. [10], integrating smart mobility solutions could lead to some improvements. The use of new technologies in the transportation system may be useful to collect, monitor, and manage real-time information: the high quantities of data collected could be processed by machine learning and artificial intelligence technologies for predictive maintenance of the process. The same data could be important in traffic and parking management, especially in large cities where traffic congestion phenomena are crucial in creating liveable cities [11]. Automated toll collection and integrated ticketing systems could be useful in promoting multimodal shifts in the context of Mobility-as-a-service [12]. Last but not least, smart mobility technologies could give advantages in surveillance and improve road user safety.

But how is it possible to evaluate how smart and sustainable a city is? According to Garau et al. [13], six variables need to be taken into account (public transport, cycle lanes, bike sharing, car sharing, private mobility support system, and public transport support system) to create a smart mobility synthetic indicator. This indicator allows for the comparison of cities to assess their smart and sustainable qualities [13].

To transform a city into a Smart City, IoT technologies need to be adopted [14]. This is achievable today through digitalization [15], maximizing the performance of digital infrastructure (such as 5G), and the recent development and widespread use of Artificial Intelligence and machine learning technologies [14] for transport automation and intelligent transport systems (ITS) [16, 17]. While improved technologies play a crucial role in collecting and analyzing a large amount of data, this shift to digitalization and modernization of infrastructures and modes of transport needs to be supported by strong Sustainable Urban Mobility Plans (SUMPs) that provide a clear long-term vision [18]. In addition to that, forecasting vehicular demand is essential to understand how these sustainable and smart mobility solutions will impact traffic scenarios and vehicular travel times. In this context, macroscopic and microscopic models, alongside simplified approaches, are often used to study traffic scenarios and promote design solutions that can meet both requirements of making a city smart and sustainable [18–21].

In this paper, a new and simplified method to forecast vehicular traffic flow based on travel times retrieved from a navigation software is proposed. The Greenshields and the BPR Models are combined with an original free flow speed prediction model to create an algorithm that could be implemented for a real-time vehicular flow estimation based on the travel times obtained from a navigation software in the context of smart mobility implementation, but also as a simple tool to be used for traffic analysis without the creation of specific and high-demanding macroscopic models.

2 Method

In this section, the methodology adopted for vehicular flow estimation is shown. Figure 1 is an example of a flow chart to graphically and logically represent the road map.

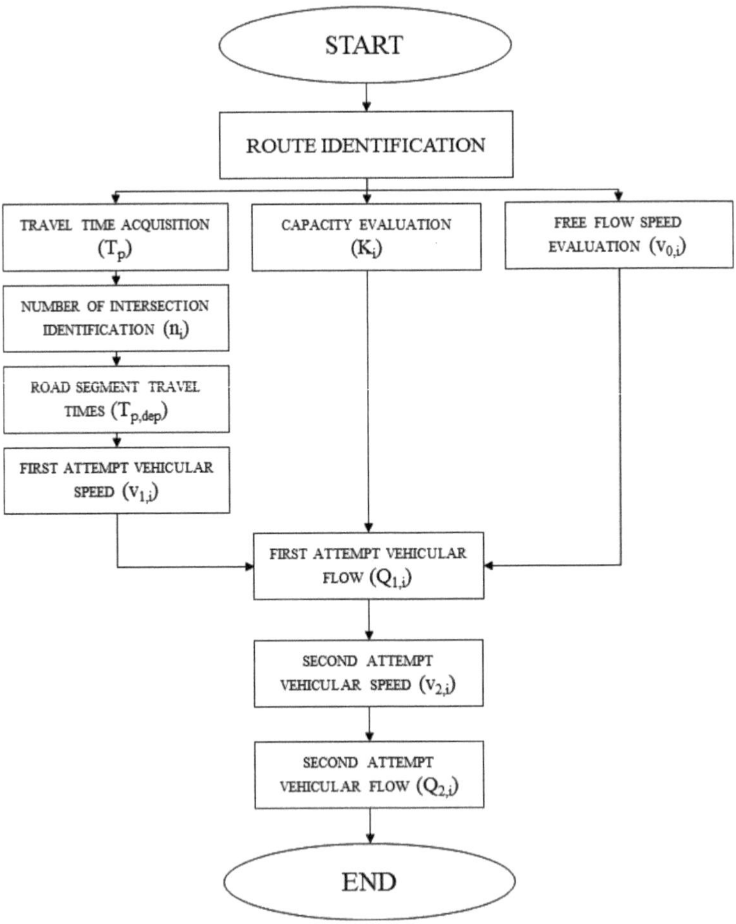

Fig. 1. Flow chart of the overall methodology.

All the steps are described in detail in the subsequent Sect. 3.1 (Greenshields Models), Sect. 3.2 (Cost Function Model), and Sect. 3.3 (Vehicular Flow Estimation Method). Sections 3.1 and 3.2 provide a theoretical framework of the two models on which the overall methodology is based, while Sect. 3.3 shows the steps of the algorithm implemented to retrieve vehicular flows. According to the flow chart in Fig. 1, the three main pillars of the methodology are represented by the Travel Times retrieved from a common navigation software, by the road section capacity (evaluated as a Saturation Flow, SF), and the Free Flow Speed (FFS).

2.1 Greenshields Model

The Greenshields model is a historical model of empirical nature based on experimental observations [22]. The traffic flow is described through the use of three physical quantities: speed, flow rate, and traffic density. Traffic flow is defined as the number of vehicles that cross the considered road section in a time interval in Eq. (1):

$$Q = \frac{n}{T} \tag{1}$$

where:

- n is the number of vehicles;
- T is the time interval.

The speed can be described in two ways: instantaneous or the average speed over time "v_t", the average of the instantaneous speeds of the vehicles passing through a section (Eulerian average speed):

$$v_t = \sum_i^n \frac{v_i}{n} \tag{2}$$

where:

- n is the number of vehicles;
- v_i is the instantaneous speed.

and the average travel speed "v_s", defined as the ratio between the length of a certain stretch and the arithmetic average of the travel times of the vehicles (Lagrangian average speed) in Eq. (3):

$$v_s = \frac{L}{\sum_i^n \frac{\Delta t_i}{n}} \tag{3}$$

where:

- n is the number of vehicles;
- Δt_i is the vehicle travel times;
- L is the length of the road.

Traffic density,ρ, is the ratio of the number of vehicles in a section at a given instant and the length of the considered road section (Eq. (4)):

$$\rho = \frac{n}{L} \tag{4}$$

where:

- n is the number of vehicles;
- L is the length of the road.

Greenshields experimentally derived a relationship between density and average travel speed. He measured the two quantities, allowing him to obtain a linear relationship between them. He measured and represented them on a Cartesian diagram. He created a trend line among all the data, obtaining the following law:

$$v_s = v_0\left(1 - \frac{\rho}{\rho_{max}}\right) \tag{5}$$

where:

- v_0 is the free flow speed;
- ρ_{max} is the maximum value of density.

The free flow speed is the speed in isolated vehicle conditions (with traffic density equal to zero). Greenshields replaced this experimentally derived formula within the fundamental traffic flow relationship, which expressed the traffic flow rate, Q, as:

$$Q = \rho \cdot v_s \tag{6}$$

obtaining another relationship that provided a link between flow and density

$$Q = \rho v_0\left(1 - \frac{\rho}{\rho_{max}}\right) \tag{7}$$

Graphically representing this relationship, he noticed a parabolic relationship between the two quantities and introduced another important measure: capacity. The latter was defined as the maximum flow with a sufficient probability of not being exceeded and marked the boundary point between two types of flow: stable flow and unstable flow.

The former indicates a system that, if perturbed, returns to its equilibrium state; the latter is a system that, if subjected to perturbation, irreversibly abandons the equilibrium condition, evolving into a chaotic condition. The problem with density was the difficulty encountered in its measurement. Greenshields rewrote the formula expressing the density as a function of speed in Eq. (8):

$$\rho = \rho_{max}\left(1 - \frac{v_s}{v_0}\right) \tag{8}$$

and the flow as a function of speed and maximum density in Eq. (9).

$$Q = v_s \rho_{max}\left(1 - \frac{v_s}{v_0}\right) \tag{9}$$

This relationship returned a parabolic graph that expressed the relationship between flow and speed. Greenshields also defined the critical speed, "*Vc*", which is the value of the speed corresponding to capacity. His model is the simplest and identifies a linear relationship between velocity V and density ρ, which results in parabolic trends $V = V(\rho)$ and $\rho = \rho(Q)$.

2.2 Cost Function Model

The elements of the network, and in particular the road segment, are associated with specific cost functions to describe quantitatively the road system. A "cost function of a road section" allows linking the average transport cost to the vehicular flows [23]. The greater the flow, the greater the travel time and the greater the cost associated with the transport. Costs could be expressed as follows:

$$C_i = C_m * L_i + VT * T_i \tag{10}$$

where:

- T_i is the travel time on the i-th road section
- C_m is the cost per unit of road length (cost of consumables, toll rate, comfort…)
- VT is the monetary value that users associate with the unit of travel time
- L_i is the road length i-th.

The component of the transport cost function that has the greatest impact on the cost is the travel time given by the sum of the physical time of travel on the road and the waiting time at the road intersections placed at the end of the road itself:

$$T_i = T_{i,r} + T_{i,w} \tag{11}$$

where:

$T_{i,r}$ is the travel time on the road.

$T_{i,w}$ is the waiting time at the intersection at the end of the road itself.

Travel times on road segment could be expressed as follow (Eq. 12) adopting the well-known US Bureau of Public Roads (BPR) formula:

$$Ti = T^0 * \left[1 + \alpha * \left(\frac{f}{K_i}\right)^{\beta}\right] \tag{12}$$

where:

Ti^0 is the Travel time of the i-th arc in free circulation, calculated as:

$$T_i^0 = 60 * \frac{L_i}{v_{p,i}} \tag{13}$$

- $v_{p,i}$ is the real road speed;
- f_i is the flow on the arch;
- α,β parameters that depend on the road characteristics;

- K_i is the road capacity, evaluated as proposed from HCM for the signalized intersections [24]:

$$K_i = N * K_i^t * f_w * f_{HV} * f_g * f_p * f_{bb} * f_a * f_{LU} * f_{LT} * f_{RT} * f_{HV} * f_{Lpb} * f_{Rpb} \quad (14)$$

where:

- K_i real capacity of the road [*veic/h*];
- $K_i{}^t$ theoric capacity of the arch [*veic/h*];
- N number of lanes;
- f_L reductive coefficient for the width of road
- f_{HV} reduction coefficient evaluated based on the percentage of heavy vehicles;
- f_g reduction coefficient for longitudinal slope;
- f_p reduction coefficient for parking activity;
- f_{bb} reduction coefficient for transit stops;
- f_a reduction coefficient for area type;
- f_{LU} reduction coefficient for lane utilization;
- f_{LT} reduction coefficient for left turns;
- f_{RT} reduction coefficient for right turns;
- f_{Lpb} reduction coefficient for interactions with pedestrians and cyclists on left turns;
- f_{Lpb} reduction coefficient for interactions with pedestrians and cyclists on right turns.

In the cost functions, in addition to the journey time, there is also the capacity, defined as the maximum hourly flow of vehicles that can reasonably pass through a point or section of a road lane during a given period, under pre-set traffic and control conditions. The HCM formulation for Capacity has been proposed since it is the most flexible in urban areas where a high amount of signalized intersections may be found.

2.3 Vehicular Flow Estimation Method

A fundamental part of this method is the estimation of the circulating flows. For the estimation of these flows, the theoretical model of Greenshields and the empirical model Bureau of Public Roads (BPR) are used, which only need the definition of the capacity and the free flow speed.

The free flow speed (v_0) was calculated using the model provided by D'Apuzzo et al. [25], by putting the flow rate term (Q) equal to zero:

$$MSS = 29.9 + 3.6L_u - 0.6S - 13.9W - 10.8H - 6.4I + 4.7X - Q(0.1027e^{-0.085L_u}) \quad (15)$$

where:

- MSS is the travel speed [km/h];
- L_u is the road width [m];
- S is the road slope [dimensionless];
- W is the tortuosity of the stretch, in values within the range [0,1];
- H is the Index of traffic disturbance caused by external factors, in values within the range [0,1];

- I is the section of road with side parking [dimensionless];
- X is a dummy variable equal to 1 if the section is paved, otherwise 0;
- Q is the flow rate [vehicle/hour].

By deleting in the Eq. 15 the flow rate term to zero, the free flow speed (v_0) is evaluated as:

$$v_0 = 29.9 + 3.6L_u - 0.6S - 13.9W - 10.8H - 6.4I + 4.7X \tag{16}$$

where:

- v_0 is the free flow speed [km/h];
- L_u is the road width [m];
- S is the road slope [dimensionless];
- W is the tortuosity of the stretch, in values within the range [0,1];
- H is the Index of traffic disturbance caused by external factors, in values within the range [0,1];
- I is the section of road with side parking [dimensionless];
- X is a dummy variable equal to 1 if the section is paved, otherwise 0.

The capacity was assimilated to the saturation flow, as calculated for the signalized intersections in the HCM LOS (Level of Service) procedure [24], as shown in Eq. 14. The main steps for determining the basic range are schematically shown below:

1. It is essential to select a route and retrieve travel times, and for this itinerary, K_i (saturation flow), v_0 (free flow speed), and CT (crossing time) must be obtained. The crossing times at intersections are initially assumed to be 20s.
2. The travel time (T_p) is obtained from a navigation software (by appropriately setting the reference time on which to set all the analyses).
3. Calculation of the travel time of the single section (T_p^*) that makes up the itinerary:

$$T_p^* = T_p - \sum_i^n CT_i \tag{17}$$

4. A constant traffic speed is calculated for the entire itinerary, which in fact represents a first attempt:

$$v_1 = \sum_i^n \frac{L_i}{T_p^*} \tag{18}$$

5. A first attempt flow rate is calculated based on the assumption of constant travel speed over the entire itinerary:

$$Q_{i(1)} = 4 * \frac{v_1}{v_{0,i}} * K_i * \left(1 - \frac{v_1}{v_{0,i}}\right) \tag{19}$$

6. With Flows and Capacity, a Saturation Rate could be obtained as follow:

$$SR = \frac{Q_{i(1)}}{K_i} \tag{20}$$

7. After that, a travel time in free flow conditions could be expressed as:

$$T^*_{p,0} = \sum_i^n \frac{L_i}{v_{0,i}} \tag{21}$$

8. An identical saturation rate is assumed for each arc that makes up the selected itinerary, therefore it is possible to estimate a new flow rate, different from that of the previous point, such as:

$$Q_{i(1)'} = K_i * \left[1 - \left(2 * \frac{T_{p0}}{T^*_{p0}} - 1\right)^2\right] \tag{22}$$

9. A new Saturation Rate could be obtained in this way:

$$SR\prime = \frac{Q_{i(1)'}}{K_i} \tag{23}$$

10. A first attempt value of the speed is then obtained according to the Greenshields model based on the value of the flow rate calculated in the previous point:

$$v\prime_{i(1)} = v_{0,i} * \left(\frac{1 + \sqrt{1 - SR\prime}}{2}\right) \tag{24}$$

11. The value of the travel time of the entire first attempt itinerary is determined as:

$$T\prime_{p(1)} = \sum_i^n \frac{L_i}{v\prime_{i(1)}} \tag{25}$$

12. A first attempt flow rate value is calculated as:

$$Q''_{i(1)} = \frac{Q_{i(1)} + Q'_{i(1)}}{2} \tag{26}$$

13. With this weighted/mean flow, another SR is evaluated:

$$SR\prime\prime = \frac{Q''_{i(1)}}{K_i} \tag{27}$$

14. A new value of speed could be retrieved, under the hypothesis of a constant SR:

$$v\prime\prime_{i(1)} = v_{0,i} * \left(\frac{1 + \sqrt{1 - SR\prime\prime}}{2}\right) \tag{28}$$

15. The value of the travel time of the entire first attempt itinerary is determined as:

$$T''_{p(1)} = \sum_{i}^{n} \frac{L_i}{v''_{i(1)}} \tag{29}$$

16. A value of k (reduction or amplification ratio) is calculated as:

$$k = \frac{T''_{p(1)}}{T'_{p(1)}} \tag{30}$$

17. It is possible to compute the new value of $v_{i\,(2)}$ as:

$$v_{i(2)} = k * v''_{i(1)} \tag{31}$$

18. Calculate the new value of $Q_{i\,(2)}$ as:

$$Q_{i(2)} = \frac{v_{i(2)}}{v_{0,i}} * K_i * \left(1 - \frac{v_{i(2)}}{v_{0,i}}\right) \tag{32}$$

19. The overall iterative procedure can be stopped when the flow rate retrieved in the previous steps is close to the one obtained in the following step, according to a defined convergence threshold, ε:

$$|Q_i - Q_{i-1}| < \varepsilon_i \tag{33}$$

3 Case Study

The proposed methodology, described in detail in the previous section, has been tested within the recent update of the Urban Traffic Plan of the city of Cassino, Lazio, Italy. Cassino is a mid-sized Italian city with approximately 35,000 residents and lies mostly on flat terrain (see Fig. 2).

Although Cassino lacks a traditional historic centre, it is still possible to identify a defined urban centre where the main commercial, administrative, and judicial activities are placed. For this reason, the city centre experiences significant vehicular flow, which intersects with equally substantial pedestrian and bicycle traffic.

Additional points of attraction include the university and a wide range of educational institutions, from kindergartens to high schools. It is therefore clear that mobility and the high demand for transportation represent major challenges for the municipality, particularly in the central areas of the city.

Fig. 2. Extension of the Municipality of Cassino.

Traffic data collection was carried out within the revision of the urban traffic plan aimed at studying the traffic impact of a new large pedestrian area, conceived to promote sustainable mobility, in the city centre of Cassino.

In the following figure (Fig. 3), the road sections where traffic counts were carried out, together with the location of the new pedestrian area, are conveniently reported.

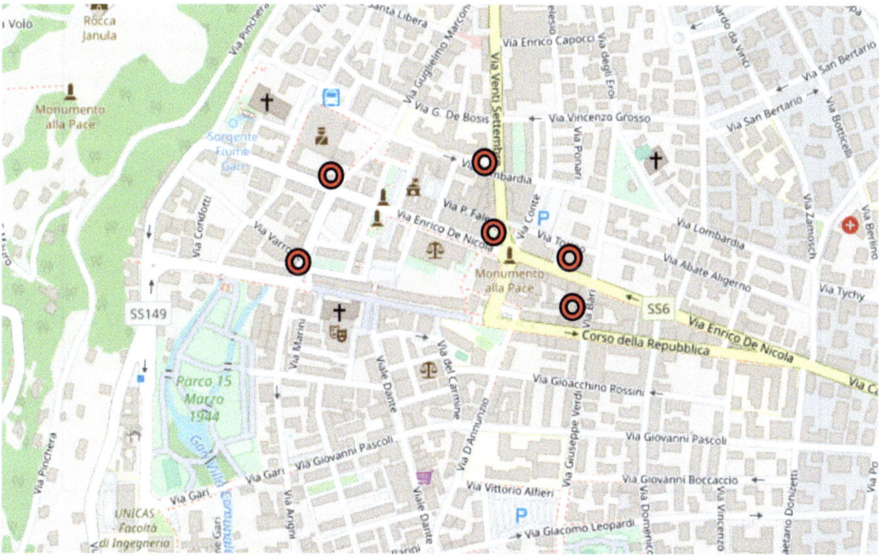

Fig. 3. City Centre of Cassino: the red dots represent the locations of traffic counts in several urban intersections, the green shaded area represents the new pedestrian area.

4 Results

After the identification of several routes and the registration of the travel times provided by a navigation software, and the analysis of the road sections that constitute the route, the vehicular flows are obtained with the method proposed in Sect. 3. The output that the algorithm provides is compared with some traffic counts collected in ten road sections. The reason is to check for model reliability. As shown in Fig. 4, the vehicular flows obtained in the first attempt with this method do not closely match the real traffic counts collected in the city of Cassino, yielding a Pearson Coefficient (P) of 0.7665. After several iterations are performed, in Fig. 5, the results are shown: in this case, the flow rate obtained at the final iteration seems to more closely match the real traffic count. It is worth noting that to improve model estimation, flows for each road section are retrieved as the mean flow by considering at least three different routes that include the specific road section.

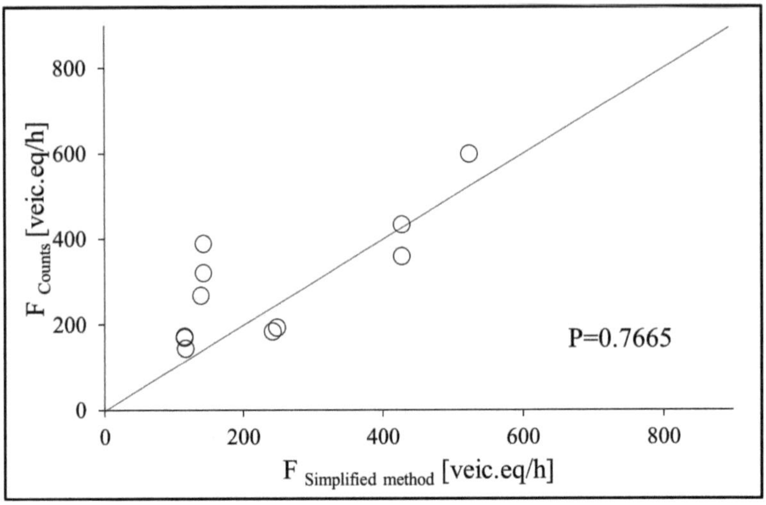

Fig. 4. Results of model validation at the first iteration: on the x-axis, $Q_{i(1)}$.

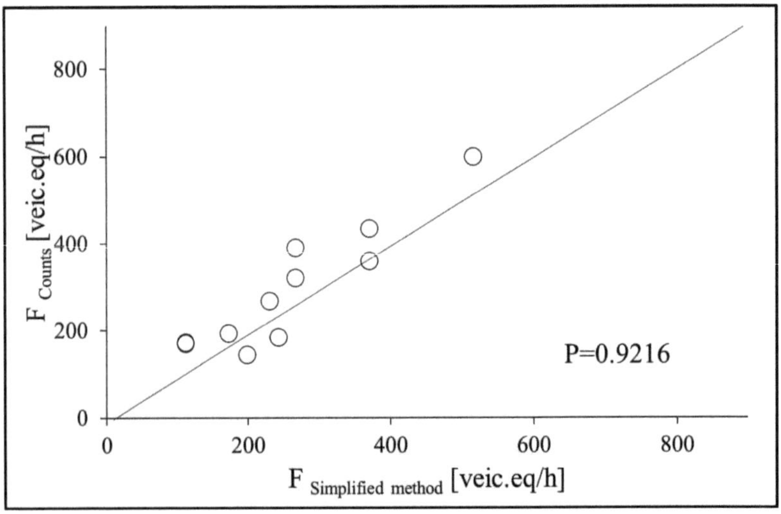

Fig. 5. Results of model validation at the end of the iterations: on the x-axis, $Q_{i(2)}$, retrieved in the final iteration.

5 Conclusions

One of the key features of smart and sustainable mobility is safety. As suggested from the analysed literature proposed in the introductory section, the spread of Artificial Intelligence and machine learning technologies, alongside transport automation and ITS, could be seen as an important prerequisite in making smart communities. All these advantages also need to be used to improve road safety. In the context of a risk-based analysis, vehicular flow evaluation is the first task to complete to evaluate vulnerable user risk and finally safety. Vehicular flows are often evaluated through highly specific and demanding macroscopic models that require high computational costs and resources. The methodology proposed here may be used on two different operating levels. The first one is higher and broader than the second one, as it aims to evaluate traffic flows in order to evaluate vulnerable road users crash risk. Such type of approach is essential, especially when professionals need to perform risk-based analysis for small communities, where the lack of crash data and the need to identify "sites with promise" to be treated is the main issue. The second lower level of the methodology aims to forecast vehicular demand on road sections based on. Within a broader context of making small communities smart and safe, this paper aims to provide a simple and alternative approach to traffic collection necessary to calibrate and validate travel demand models. The proposed methods allow users to evaluate vehicular flows starting with travel times retrieved from a navigation software by combining the Greenshields, BPR model, and a free flow speed model, *ad hoc* developed. As it is possible to notice in the results section, the proposed model has been validated, and the correlation coefficient (Pearson coefficient) shows quite good results.

As already noticed, this methodology could be easily implemented within road safety analysis in small communities (where there is not much available data), but also to forecast actual vehicular traffic flows. As it emerges from Fig. 5, the overall prediction performance seems somewhat satisfactory, although a slight overestimate for most of the experimental points may be identified. The time lag between the traffic count collection and the travel time evaluation by means of navigation software may have impacted the actual agreement between the observed and predicted traffic flow data.

Within a future development of this methodology, in order to promote generalization of the results, it is planned to extend the comparisons of estimated with actual traffic flows not only to more road sections within the city, but also to different urban contexts and to other small communities.

Acknowledgments. This study was carried out within the MOST – Sustainable Mobility Center and received funding from the European Union Next-GenerationEU (PIANO NAZIONALE DI RIPRESA E RESILIENZA (PNRR) – MISSIONE 4 COMPONENTE 2, INVESTIMENTO 1.4 – D.D. 1033 17/06/2022, CN00000023). This manuscript reflects only the authors' views and opinions, neither the European Union nor the European Commission can be considered responsible for them.

References

1. World Population Prospects. https://population.un.org/wpp/, Accessed 7 July 2025
2. Paiva, S., Ahad, M.A., Tripathi, G., Feroz, N., Casalino, G.: Enabling technologies for urban smart mobility: recent trends, opportunities and challenges. Sensors. **21**, 2143 (2021)
3. Gouveia, J.P., Seixas, J., Giannakidis, G.: Smart city energy planning: integrating data and tools. In: Proceedings of the 25th International Conference Companion on World Wide Web, pp. 345–350 (2016)
4. Chourabi, H., et al.: Understanding smart cities: an integrative framework. In: 2012 45th Hawaii International Conference on System Sciences, pp. 2289–2297. IEEE (2012)

5. Lai, C.S., et al.: A review of technical standards for smart cities. Clean Technol **2**, 290–310 (2020)
6. Ghazal, T.M., et al.: IoT for smart cities: Machine learning approaches in smart healthcare—a review. Future Internet. **13**, 218 (2021)
7. Maldonado Silveira Alonso Munhoz, P.A., et al.: Smart mobility: The main drivers for increasing the intelligence of urban mobility. Sustainability **12**, 10675 (2020)
8. Flügge, B.: Smart mobility–connecting everyone: Trends, concepts and best practices. Springer (2017)
9. Smart cities - European Commission. https://commission.europa.eu/eu-regional-and-urban-development/topics/cities-and-urban-development/city-initiatives/smart-cities_en, Accessed 7 July 2025
10. Benevolo, C., Dameri, R.P., D'auria, B.: Smart mobility in smart city: action taxonomy, ICT intensity and public benefits. In: Empowering Organizations: Enabling Platforms and Artefacts, pp. 13–28. Springer (2015)
11. Yassin, H.H.: Livable city: an approach to pedestrianization through tactical urbanism. Alex. Eng. J. **58**, 251–259 (2019)
12. Mobility as a Service - EURObizEURObiz. https://www.eurobiz.com.cn/mobility-as-a-service/, Accessed 7 July 2025
13. Pinna, F., Masala, F., Garau, C.: Urban policies and mobility trends in Italian smart cities. Sustainability. **9**, 494 (2017)
14. Allam, Z., Dhunny, Z.A.: On big data, artificial intelligence and smart cities. Cities **89**, 80–91 (2019)
15. Alvarez, R.: The relevance of informational infrastructures in future cities. field actions science reports. J. Field Actions, 12–15 (2017)
16. Directive - 2010/40 - EN - EUR-Lex. https://eur-lex.europa.eu/eli/dir/2010/40/oj/eng, Accessed 7 July 2025
17. Sustainable and Smart Mobility Strategy – European transport on track for the future. https://www.2zeroemission.eu/mediaroom/sustainable-and-smart-mobility-strategy-european-transport-on-track-for-the-future/, Accessed 7 July 2025
18. D'Apuzzo, M., Cappelli, G., Naranni, S., Pilla, G.F., Nicolosi, V.: Fostering urban soft mobility by means of pedestrian areas: tentative criteria for evaluation of impacts on traffic. In: International Conference on Computational Science and Its Applications, pp. 39–56. Springer (2024)
19. D'Apuzzo, M., Cappelli, G., Buzzi, S., Nicolosi, V.: Smart Urban Mobility Management project: a concrete step towards more sustainable and connected communities. In: 2022 Second International Conference on Sustainable Mobility Applications, Renewables and Technology (SMART), pp. 1–8. IEEE (2022)
20. D'Apuzzo, M., Evangelisti, A., Cappelli, G., Naranni, S., Buzzi, S., Nicolosi, V.: Towards an urban smart mobility: preliminary results of an experimental investigation in artena. In: International Conference on Computational Science and Its Applications, pp. 179–193. Springer (2023). https://doi.org/10.1007/978-3-031-37123-3_15
21. D'Apuzzo, M., Cappelli, G., Santilli, D., Evangelisti, A.: Sustainable rehabilitation of urban city: a new design method for pedestrian areas. In: The RE-BUILT Multiplier Event held in the form of the International Conference on Knowledge Transfer on Sustainable Rehabilitation and Risk Management in the Built Environment. pp. 315–325. Springer (2021). https://doi.org/10.1007/978-3-031-43455-6_29
22. Kessels, F., Kessels, R., Rauscher, R.: Traffic flow modelling. Springer (2019)
23. Zhang, J., Liu, M., Zhou, B.: Analytical model for travel time-based BPR function with demand fluctuation and capacity degradation. Math. Probl. Eng. **2019**, 5916479 (2019)

24. Reilly, W.: Highway capacity manual 2000. Tr News.(1997)
25. D'Apuzzo, M., et al.: Toward the development of a hybrid approach to speed estimation in urban and rural areas. Traffic Inj. Prev. **22**, 449–454 (2021)

The Impact of Road Conditions on Energy Consumption in the Electric Drive System of a Passenger Car – Experimental Analysis in the RDE Cycle

Maciej Sproch, Mariusz Graba(✉), Krystian Hennek, and Jarosław Mamala

Faculty of Mechanical Engineering, Opole University of Technology, Opole, Poland
m.sproch@student.po.edu.pl, {m.graba,k.hennek,j.mamala}@po.edu.pl

Abstract. This paper presents an analysis of the actual electric energy consumption of a battery electric vehicle (BEV) passenger car under real-world operating conditions, using a road test performed in accordance with the Real Driving Emissions (RDE) procedure. The aim of the study is to determine the impact of urban, rural and motorway traffic on the energy efficiency of BEVs, expressed in mileage-based electric energy consumption. In the experimental studies, using the case study method, various methods of measuring mileage-based electric energy consumption were compared, including an analysis of data read from the on-board diagnostic network and calculations based on the capacity of the traction battery and the amount of electric energy consumed from the power grid, related to the mileage covered by the vehicle, as well as data obtained from the Global Positioning System (GPS). In the study three types of routes were analysed: urban (characterised by frequent changes in traffic phases separated by periodic stops), rural (more fluid, with a higher average speed) and motorway (with predominantly high speed and aerodynamic drag). For each of these scenarios, the average electric energy consumption was determined and the average travel speeds were calculated, which were used to compare the energy consumption of an electric passenger car under different traffic conditions Conducting the analysis made it possible to identify differences in energy consumption between traffic types, confirming the influence of the speed profile on the RDE teste results. A proposal was also presented to modify the RDE cycle procedure to include correction factors for different types of routes, allowing for a more accurate estimation of actual electric energy consumption in relation to type-approval data. The research outcomes can be applied to improve the accuracy of range estimation algorithms, support the development of more representative driving cycles, and guide policymakers in refining type-approval procedures by introducing correction factors for different traffic conditions. In the article the need for further research on the impact of external conditions and individual driver habits on the energy consumption of electric passenger cars was emphasized.

Keywords: Worldwide Harmonized Light-Duty Vehicles Test Procedure · Real Driving Emissions cycle · Battery Electric Vehicle · electric energy consumption · electric energy consumption measurement · road conditions · road test guidelines

E. Macioszek et al. (Eds.): TSTP 2025, LNNS 1769, pp. 138–153, 2026.
https://doi.org/10.1007/978-3-032-14094-4_10

1 Introduction

With the development of electromobility and growing awareness of the need to reduce greenhouse gas emissions, electric vehicles (EVs) are becoming an increasingly important part of a sustainable transport system [1]. The issue of reliably assessing a vehicle's energy consumption under real operating conditions is becoming increasingly important, as there is a discrepancy between the range declared by car manufacturers and their actual performance. In their article, Weiss et al. compare the differences between the declared and actual range. The authors propose the use of energy labels on a scale of A to G, which would cover energy consumption, battery capacity and range. This would allow consumers to compare vehicles not only in terms of efficiency but also in terms of functionality [2]. From the point of view of research, the authors point out that a key aspect of assessing the suitability of electric passenger cars is the indicator of electric energy consumption per kilometre, and above all whether it is precisely defined for real traffic conditions. In many scientific publications, trade journals and technical websites describing vehicle specifications, there are discrepancies between the specific mileage electric energy consumption in the car's technical specifications and the result obtained for actual consumption in road driving conditions. Over the last 7 years (2017–2024), many changes have been made in the field of road condition simulation testing, including the introduction of the new WLTP (Worldwide Harmonised Light Vehicles Test Procedure), driving cycle. These changes mainly concern the requirements for exhaust emission testing of drive systems. Since September 2019, a new RDE road cycle has also been introduced, which has become mandatory for all new passenger cars and light commercial vehicles in the European Union [3, 4]. Unlike traditional laboratory driving cycles such as New European Driving Cycle (NEDC) or the new WLTP, the new RDE cycle requires the measurement of fuel consumption and harmful emissions under real-world conditions, taking into account driving style variability, traffic intensity, specific route profile and weather conditions over a sufficiently long road section [3, 5–7].

Scientific studies clearly indicate significant discrepancies between the results obtained under homologation conditions (e.g. in the WLTP cycle) and the actual fuel and electric energy consumption of vehicles – both with conventional hybrid and electric powertrains [8]. These differences result, among other things, from simplified test conditions that do not fully reflect the variability of actual traffic conditions, driving styles or terrain [9]. Literature reviews confirm that factors such as driving style, ambient temperature, vehicle weight [10], route profile and changes in driving speed [11], have a significant impact on the electric energy consumption of electric powertrains [12–16]. These types of drive systems are still being tested in laboratory conditions on a chassis dynamometer. Under such conditions, it is difficult to reproduce the variability of the drive system load resulting from road traffic conditions. Equally important in determining energy consumption are the individual preferences of the driver, which to a greater or lesser extent affect the total energy expenditure of the vehicle. With regard to the driver's style of driving an electric car, aggressive driving in urban conditions can increase energy consumption by 16% [12] and increasing the vehicle's weight by 100 kg results in an increase in consumption of approximately 0.34 kWh/100 km [16]. The impact of low ambient temperatures on the range of an electric car is significant at ambient temperatures below 0 °C. The range and may decrease by 31.9% due to the need to heat the vehicle interior and 21.7% due to the limited recuperation system [17], while the differences between the results obtained in the laboratory and the actual results may be as high as 25% [16]. Therefore, increasing importance is being attached to the development and implementation of appropriate correction factors that allow for a more reliable estimation of electric energy consumption in various road conditions, especially in urban and motorway traffic [18].

There are many publications describing the methodology for calculating or obtaining the mileage electric energy consumption in real conditions [19–21]. In practice, when analysing data collected from an on-board diagnostic network data recording device, electric energy consumption is determined on the basis of energy meter readings or the difference in battery state of charge (SOC) at a known nominal voltage and energy capacity required to power the electric motor as the drive source. Alternatively, charging data is used, but in this case the results also include losses resulting, among other things, from the efficiency of the charger and the operation of vehicle systems that maintain the traction battery in conditions of safe and efficient electric energy consumption. In road tests, the energy consumption value refers to a specific section of the route – urban, suburban, motorway or full distance – and the data obtained is subjected to time-speed segmentation in order to identify the impact of traffic conditions. In addition, comparative analyses are used in relation to homologation data, which allows the operational efficiency of a given vehicle to be assessed under specific test conditions.

The discrepancy between the mileage-based electric energy consumption declared on the basis of homologation tests and actual consumption under operating conditions is a significant problem in assessing the efficiency of electric vehicles. Type-approval data, most often based on the results of laboratory tests carried out in accordance with the WLTP present average and optimistic energy consumption values that are rarely achieved under real operating conditions. Road tests show that actual electric energy consumption can be 15–40% higher than homologation values, depending on weather conditions, terrain topography, driving style, use of on-board systems (e.g. heating, air conditioning) and vehicle load [9, 14, 22]. For example, the EV365 (Electric Vehicle 365) project recorded an average energy consumption of 11.5 kWh/100 km in urban use, while the homologation value for the tested model (Škoda Citigo-e iV) was approximately 12.8 kWh/100 km, which could suggest a favourable convergence – however, in winter conditions or on suburban routes, actual consumption reached 14–16 kWh/100 km [23]. The literature also emphasises the importance of climatic conditions – in winter, electric vehicles consume significantly more energy for cabin heating, which can lead to a decrease in effective range of up to 30–40% compared to the declared values [9, 22]. Such differences are significant from the point of view of both end users (e.g. travel planning) and energy and transport policy. As a result, road tests in RDE cycles and the analysis of real-world fleet operation data (On-Board Fuel Consumption Monitoring (OBFCM), on-board telemetry) are playing an increasingly important role as a supplement to standard type-approval procedures [24].

Taking into account the literature review and the discrepancy regarding the analysis of energy consumption for electric cars, the aim of this article is to analyse the impact of road conditions on the mileage energy consumption of an electric passenger car drive system, using the RDE test procedure as representative of combustion engine drive systems.

Particular attention was paid to various methods of measuring the energy consumption of electric vehicles. As part of own research for the developed RDE cycle in the area of the city of Opole, which includes three types of routes: urban, rural and motorway, the average energy consumption values were determined, and the values for the average energy consumption during the entire cycle were also referenced. The weights of the impact of individual route types on the average electric energy consumption values were determined in relation to the lengths of individual sections of the driving cycle and average driving speeds. The results of the study are important for both vehicle manufacturers and users in order to minimise discrepancies between technical data and values resulting from actual road conditions.

2 Research Methodology

The analysis of the methodology for developing the RDE (Real Driving Emission) cycle is based on a set of guidelines, according to which the research should be conducted in road conditions, but without a strictly imposed speed profile. The guidelines include, among other things, the length of individual route types, average driving speeds and a limited duration of the test drive. Currently, there are no requirements (or International Organization for Standardization (ISO) standards) that directly refer to a special test procedure for vehicles equipped with electric drive systems. Therefore, for the above-mentioned testing purpose, a case study method was used, including both experimental measurements and analysis of the determined electric energy consumption, thus pointing to a research gap in the form of a lack of standardisation of the testing procedure and discrepancies in the results obtained in non-laboratory tests.

2.1 Research Object

The research object was a passenger car, described in the literature and also used in the EV365 study, equipped with an automated electric drive system (BEV) by Škoda, model Citigo-e IV generation, manufactured in 2019. The vehicle has an electric drive powered by a set of lithium-ion batteries with a net capacity of 32.3 kWh (total capacity: 36.8 kWh), allowing it to travel 265 km in the WLTP cycle. The BEV's kerb weight is 1235 kg, with 248 kg attributable to the traction battery mounted in the chassis. The technical data is summarized in the Fig. 1.

The electric vehicle selected for road testing is one of the A-segment passenger cars, i.e. the smallest passenger cars, and is a model characterised by compact dimensions and urban use. From the point of view of road testing in real road conditions, it is also a design that requires familiarisation in terms of control. Therefore, in order to drive the vehicle in a fully conscious, efficient and smooth manner, it is necessary to first familiarise oneself with its control system and then attempt to achieve similar results in terms of electric energy consumption on the selected test route. In the case of the tests described, the route was covered by one driver, and the total distance travelled was approximately 835 km. Thanks to this, and by maintaining identical vehicle settings, it was possible to maintain a consistent driving style and reliably compare energy results. The total weight of the driver and vehicle during the tests was 1335 kg, which is the

TECHNICAL DATA	VALUE
Ross battery capacity	36.8 kWh
Net battery capacity	32.3 kWh
Maximum speed	130 km/h
Average energy consumption WLTP	12.6 kWh/100km
WLTP range	265 km
Engine power	61 kW
Torque	212 Nm
Curb weight	1235 kg

Fig. 1. Characteristics of a car with an electric drive system

sum of the vehicle's curb weight (1235 kg) and the driver's weight (100 kg). Despite the urban purpose of the tested vehicle, the test drives included a mixed cycle, characterised by different traffic resistance, road conditions, dynamics and average driving speeds. The authors' approach makes it consistent with the assumptions of real road tests.

2.2 Research Tools

It is crucial for real-world road tests to ensure that traction parameters and data obtained from the drive system are recorded. The use of an autonomous measurement system integrated with a the Global Positioning System (GPS) enabled the recording of dynamic vehicle parameters such as speed, acceleration, distance travelled, changes in altitude, as well as accurate location reference to the terrain map. By ensuring an identical sampling frequency (up to 100 Hz), it is possible to compare measurement data, e.g. to create a speed profile in relation to time or distance travelled, as shown in Fig. 2.

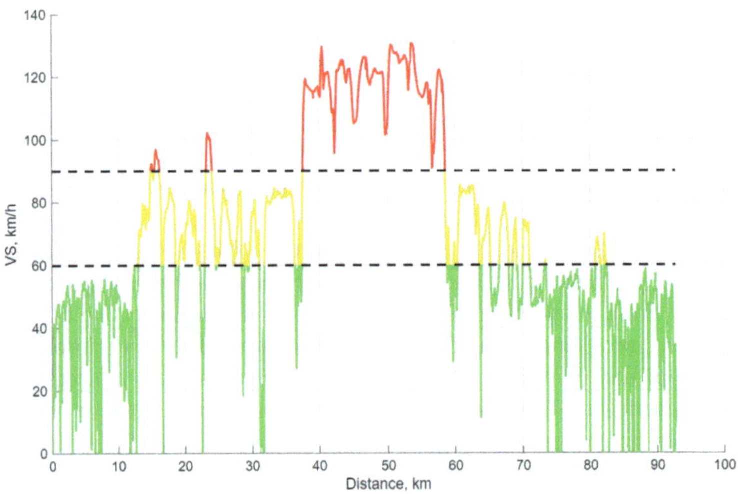

Fig. 2. Recorded speed profile in relation to distance travelled for an example research test, VS – vehicle speed

A proprietary measurement system based on the LabView programming environment was used to record the traction, kinematic and energy parameters of the car's drive system. It was used to obtain measurement data from the on-board diagnostic system by developing appropriate identifiers in the UDS (Unified Diagnostic Services) system (Request-Responses), characterising the test vehicle's data transmission methodology. The following parameters were recorded and saved for the electric motor itself: power, rotor speed, electrical current and voltage, as well as high-voltage battery indicators: capacity, voltage and high-voltage (HV) current. Also, additional data were gathered the position of the accelerator pedal and the total mileage of the vehicle. Taking appropriate data configurations, e.g. only for the electric motor, it is possible to calculate the mileage-based electric energy consumption without taking into account the electric energy needed to power additional devices.

In order to determine the mileage-based electric energy consumption based on the amount of energy consumed from the grid, a charger with a data recording function was used. At the same time, the use of the charger shaped the way the tests were conducted, which consisted of charging the traction battery to 100% capacity each time, performing a test run, and recharging the battery with data recording and storage during the charging of the traction battery.

2.3 Developed RDE Cycle

The road cycle was designed in accordance with the requirements of the RDE procedure and included three main types of traffic conditions: urban traffic (frequent stops, acceleration and braking), rural traffic (smooth driving at a higher average speed) and motorway traffic (high speed, aerodynamic drag). The test route was a proprietary road cycle, most of which runs within the administrative boundaries of the city of Opole. Due to the motorway section of traffic required for the tests, the route was extended to include a section of the A4 motorway starting at the Opole-West entrance and ending at the Opole-Racibórz-Krapkowice exit.

The average distance of the entire route was 92.6 km, with the start and end of the route being the premises of the Faculty of Mechanical Engineering of the Opole University of Technology. The average speed achieved over the entire distance ranged between 40 and 60 km/h. The remaining road requirements for the RDE cycle in relation to the area of the City of Opole are summarised in Table 1, while the graphical course of the route for the designated cycle is presented in Fig. 3.

An important element of the developed RDE road cycle is also the topography of the terrain along the entire route due to the elevation of the road. The values obtained in relation to the total length of the route made it possible to verify whether this requirement for the RDE cycle was met. The guidelines stipulate that the maximum elevation should not exceed 1200 m per 100 km (i.e. 12 m/km). In the analysed case, this value was 4.9 m/km, which means that the route was within the limits acceptable for the RDE cycle and can be considered representative in terms of topography.

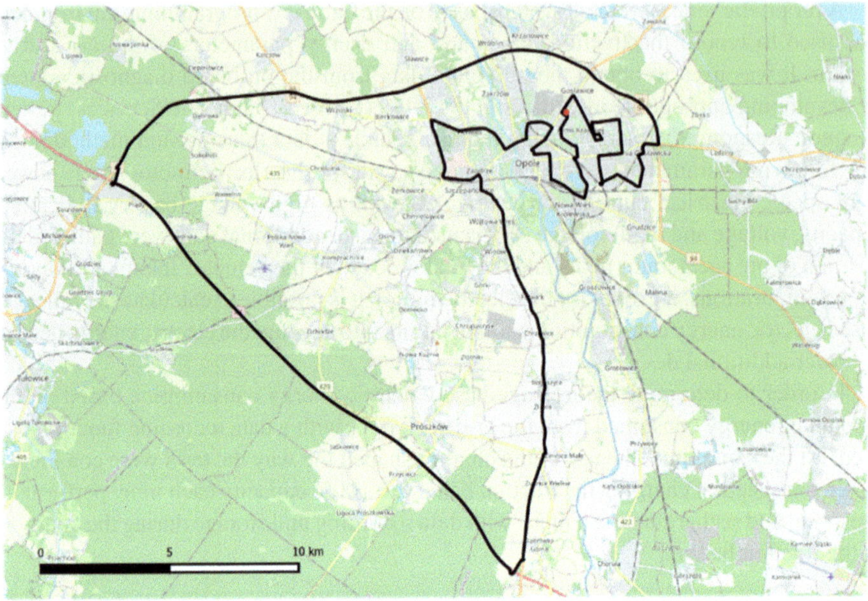

Fig. 3. Route for the developed RDE road cycle

Table 1. Road requirements for the proprietary RDE cycle running through the city of Opole.

PARAMETER	VALUE	RDE TEST CONDITIONS
Route loop area	**The city of Opole and the surrounding area**	Driving area variability: - urban area - rural area - motorway area
Total driving time	**114 min**	90–120 min
Total distance	**92.6 km**	>48 km
Average driving speed in urban areas	**31.0 km/h**	15 km/h–40 km/h
Average driving speed in rural areas	**74.9 km/h**	60 km/h–90 km/h
Average speed in motorway areas	**113.2 km/h** **11.5 min**	> 90 km/h (>100 km/h for 5 min)
Distance in urban areas	**39.6 km**	>16 km
Distance in rural areas	**28.7 km**	>16 km
Distance in motorway areas	**26.3 km**	>16 km
Percentage of urban areas	**42.8%**	29%–44%
Percentage of rural areas	**31.0%**	23%–43%
Percentage of motorway areas	**26.1%**	23%–43%
Percentage of time spent stationary in urban areas	**27%**	6%–30%

2.4 Methods of Measuring Energy Consumption

For the purposes of own research, two independent methods of measuring electric energy consumption were used, which are summarised in the Table 2, together with a description of how they were implemented.

Table 2. Methods of measuring electric energy consumption.

No	METHOD	IMPLEMENTATION
1	Recording of electric energy consumed from the power grid	A charger with a function for recording data on electric energy consumed from the power grid was used to power the vehicle's traction battery
2	Data recording from the on-board diagnostic system BEV	The average electric energy consumption per 100 km, distance travelled and driving speed were recorded

When reading from the on-board diagnostic network, the average electric energy consumption was recorded at 11.25 kWh/100 km. All test drives were carried out with identical drive system settings, the air conditioning system switched off, and the vehicle driven by a single driver.

In Fig. 4 the total average electric energy consumption in the road cycle recorded by the methods were compared.

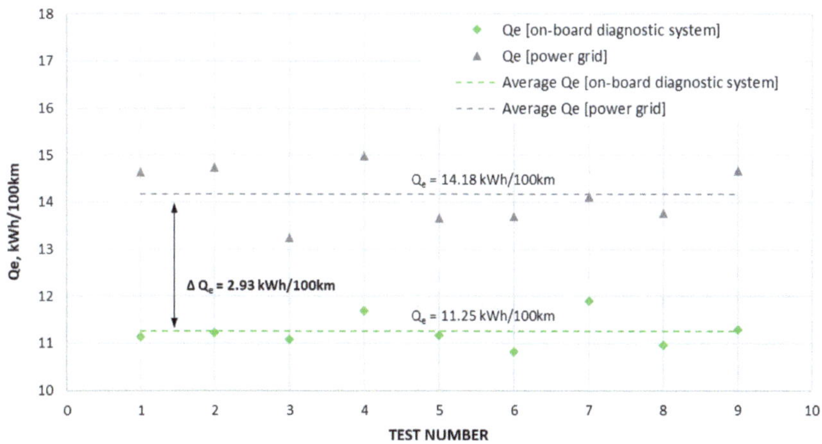

Fig. 4. Differences in electricity consumption values between the value taken from the charger and the value recorded by the on-board diagnostic system.

The reference point for the data expressing the amount of electricity actually drawn from the power grid (13.13 kWh) was the average electricity consumption obtained while driving, based on data from the on-board diagnostic system. After relating the recorded amount of energy from the grid to the full cycle distance, the gross electric energy consumption was determined to be 14.18 kWh/100 km. The absolute difference is 2.94 kWh/100 km. This value is nearly 21% higher and is mainly due to additional processes accompanying charging, resulting from maintaining the appropriate temperature of the traction battery (cooling/heating systems), energy losses from the limited efficiency of the charger, as well as heat losses and Alternating Current (AC)/ Direct Current (DC) voltage conversion losses. This value is significantly higher than the values given in the literature, which mainly report the average charging efficiency rather than the efficiency resulting from the conversion of electric energy into driving force at the wheels [25–28].

The net efficiency of the vehicle's propulsion system was calculated by converting the energy drawn from the traction battery (based on its state-of-charge change) into mileage-based energy consumption per 100 km, and comparing this value with the result obtained from the on-board diagnostic data. The average difference was found to be 2.16 kWh/100 km, as shown in Fig. 5.

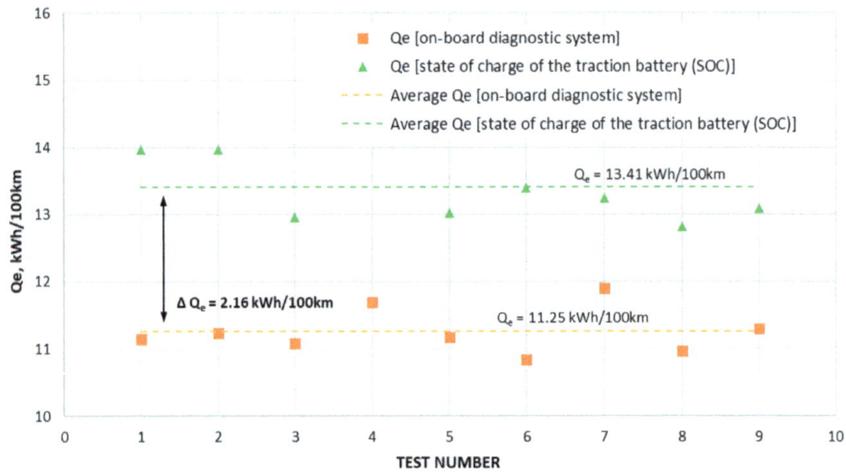

Fig. 5. . Differences in the values of runtime electricity consumption between the value taken from the battery and the value recorded by the on-board diagnostic system.

The difference is 16.1%, which indicates that the mean efficiency of the drive system is 83.9%. The EV365 study also recorded a similar average value of 11.5 kWh/100 km. The convergence of the results confirms that the average consumption values obtained are representative of small BEVs in real-world conditions [23].

3 Research Process and Data Analysis

The subject of the analysis is research conducted on the developed RDE road cycle presented in Fig. 2, which was determined in accordance with RDE guidelines and the classification described in the literature in relation to the speed profile. Although European Union (EU) regulations, such as Regulation (EU) 2017/1151, defines the classification of driving areas solely on the basis of speed ranges, this study also examined an alternative approach that also takes into account local road conditions, traffic characteristics and travel structure, which allows for a better reflection of actual operating conditions and their impact on the energy consumption of electric vehicle traffic. Classification according to road conditions means that momentary speed violations resulting, for example, from overtaking manoeuvres (details A and B) or momentary deceleration of the vehicle in the cycle do not change its road traffic conditions, as shown in Fig. 6.

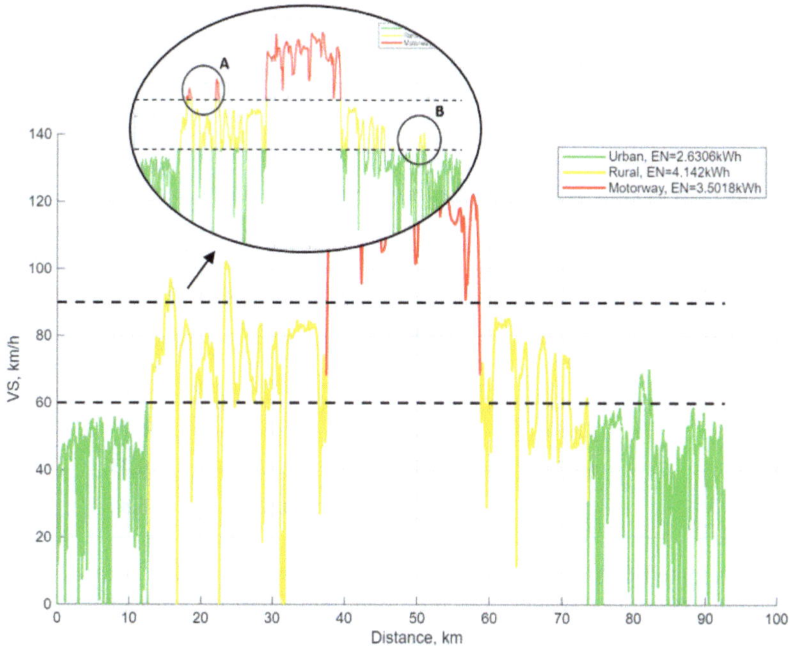

Fig. 6. Example speed profile for a selected test run in relation to road conditions

In this way, for the adopted speed profile classification, traction and energy parameters were determined in relation to the driving speed in the cycle and in relation to road conditions, as summarised in Tables 3 and 4 respectively.

These parameters are coherent with the accepted methodology for classifying the conditions in which a car travels on the road, described in numerous studies [7, 12, 13, 29–31], for which the parameters are consistent with those specified for the RDE cycle in terms of length and average travel speeds. When analysing this speed profile, local instantaneous speed exceedances can be noted, which affect the traction and energy parameters of the tested RDE cycle in terms of its classification into urban, rural and motorway conditions, as shown in Table 4.

Table 3. Summary of average traction and energy parameters of the speed profile –classification by speed.

Driving area	Traction and energy parameters			
	Speed [km/h]	Distance [km]	Time [min]	Energy Consumption [kWh/100 km]
URBAN	**31.05**	**39.61**	**77.24**	**8.62**
RURAL	**74.58**	**28.77**	**23.15**	**10.60**
MOTORWAY	**112.82**	**24.22**	**12.89**	**16.37**

Table 4. Summary of average traction and energy parameters of the speed profile – classification according to road conditions.

DRIVING AREA	TRACTION AND ENERGY PARAMETERS			
	Speed [km/h]	Distance [km]	Time [min]	Energy Consumption [kWh/100 km]
URBAN	**29.88**	**31.64**	**64.24**	**8.67**
RURAL	**63.29**	**39.72**	**37.79**	**10.33**
MOTORWAY	**113.53**	**21.24**	**11.26**	**16.84**

For the adopted classification of the RDE cycle speed profile, percentage shares and weighting factors of specific driving conditions were determined for each area: urban, rural, and motorway, as presented in Tables 5 and 6. The presented data represent the average values from nine RDE driving tests. The mileage-based electric energy consumption (Q_e), expressed in kWh/100 km, refers to the net energy drawn from the traction battery, determined on the basis of data recorded by the vehicle's on-board diagnostic system. The main parameters recorded were the voltage and current of the high-voltage (HV) battery. This approach made it possible to focus the study on the actual energy consumed directly by the traction battery during driving.

Table 5. Summary of average traction and energy indicators for the RDE cycle speed profile – classification based on speed.

TEST NO	ANALYSIS OF TRACTION AND ENERGY PARAMETERS							
	Speed	Distance	Time	Q_e	PARAMETERS	DRIVING AREA		
	[km/h]	[km]	[min]	[kWh/100km]		URBAN	RURAL	MOTORWAY
AVERAGE	**49.27**	**92.60**	**113.29**	**11.25**	**v [km/h]**	62.91%	152.08%	229.83%
					l [km]	42.78%	31.07%	26.15%
					t [s]	68.04%	20.54%	11.43%
					Q_e [kWh/100km]	76.55%	94.17%	145.52%

Table 6. Summary of average traction and energy indicators for the RDE cycle speed profile – classification based on road conditions.

TEST NO	ANALYSIS OF TRACTION AND ENERGY PARAMETERS							
	Speed	Distance	Time	Q_e	PARAMETERS	DRIVING AREA		
	[km/h]	[km]	[min]	[kWh/100km]		URBAN	RURAL	MOTORWAY
AVERAGE	**49.27**	**92.60**	**113.29**	**11.25**	**v [km/h]**	60.51%	128.84%	230.87%
					l [km]	34.17%	42.89%	22.93%
					t [s]	56.59%	33.46%	9.95%
					Q_e [kWh/100km]	77.04%	91.74%	149.67%

For the data presented in Tables 5 and 6, standard deviations from the mean values were determined for the various speed profile classification criteria analysed. The results are summarised in Table 7.

The standard deviation values (Table 7) indicate that both classifications used in the final assessment yielded identical standard deviations for total electric energy consumption in the full-cycle evaluation. However, when considering the specific driving areas in which the vehicle operated during the test, noticeable differences emerge. Taking into account the weighting of electric energy consumption across the different driving segments, the resulting changes are presented in Table 8.

Table 7. Summary of standard deviations for the adopted classification.

DIVISION	STANDARD DEVIATIONS (±)							
	Speed	Distance	Time	Q_e	PARAMETERS	DRIVING AREA		
	[km/h]	[km]	[min]	[kWh/100km]		URBAN	RURAL	MOTORWAY
Classification by speed (Table 5)	±3.31	±0.15	±7.920	±0.32	**v [km/h]**	±2.43%	±11.65%	±14.54%
					l [km]	±1.71%	±1.60%	±1.45%
					t [s]	±2.42%	±1.92%	±1.03%
					Q_e [kWh/100km]	±1.96%	±3.45%	±7.85%
Classification according to road conditions (Table 6)	±3.31	±0.15	±7.919	±0.32	**v [km/h]**	±2.86%	±9.80%	±11.38%
					l [km]	±0.12%	±0.08%	±0.04%
					t [s]	±2.62%	±2.48%	±0.49%
					Q_e [kWh/100km]	±2.12%	±4.60%	±8.24%

Table 8. Summary of the calculation of electric energy consumption per kilometre for the adopted classification and traffic conditions

CLASSIFICATION	PARAMETERS	DRIVING AREA		
		URBAN	**RURAL**	**MOTORWAY**
Subdivision according to speed (Table 5)	**Q_e [kWh/100km]**	8.62	10.60	16.37
Subdivision according to road conditions (Table 6)	**Q_e [kWh/100km]**	8.67	10.33	16.84
Relative difference	%	-0.64	2.69	-2.80

The data presented in Table 8 highlight the sensitivity of electric energy consumption measurements to the adopted classification method. Changes in road conditions, particularly the variation in the driving distance within urban areas, led to a decrease in the urban segment's share of the RDE cycle distance from 42% to 34%. Despite this reduction, no change in the energy consumption rate for urban driving was observed. This is due to the fact that the average distance covered in this segment remained relatively constant. The situation differs in rural and motorway segments. In the rural area, the distance increased from 31% to 42.9%, while the average speed decreased by 23.2%. This reduction in speed contributed to a 2.69% decrease in relative energy consumption per kilometre. Conversely, in the motorway segment, the share of the distance decreased from 26.15% to 22.93%, but the average speed increased by 1.04%. This speed increase resulted in a 2.8% rise in mileage-based electric energy consumption by the drive system.

4 Conclusion

Road conditions are an important factor influencing the average energy consumption values obtained for the basic means of transport, primarily due to the average speed achieved for a specific driving profile. The results obtained are highly repeatable in relation to the full RDE road cycle – the standard deviations are similar and do not exceed 3% of the average values, which confirms the stability of the measurements of average energy consumption for electric drive, regardless of the method used to classify driving conditions. This is beneficial for test comparability. The lowest energy consumption was recorded in urban traffic (approx. 8.6 kWh/100 km), higher in rural traffic (approx. 10.3 – 10.6 kWh/100 km), and the highest on the motorway (approx. 16.4 – 16.8 kWh/100 km). The values of energy consumption per kilometer are similar to the results described by other authors in the publication [23], but when considering the area in which the car was driven (urban, rural, motorway), differences can be observed. The decisive factor is the average speed of the tested car in the driving cycle. The tests were carried out by a single driver who was informed about the Hawthorne effect, while the authors are aware of the vehicle control process used in the tests. Due to the duration and length of the route for the selected RDE cycle, the driver focuses primarily on road traffic conditions and the environment. In WLTP homologation tests, the driver's influence is limited because their actions are dictated by the speed profile, and yet the achieved energy consumption is 12.6 kWh/100 km, which differs from that obtained in road conditions in the RDE cycle, where the average value is 11.25 kWh/100 km. However, in relation to the value resulting from the state of charge (SOC) of the traction battery and the distance traveled in the RDE cycle, this value is higher and amounts to 13.41 kWh/100 km. For the tested car, this is an acceptable value, which confirms the influence of other factors characteristic of vehicle operation in real conditions. When comparing the mileage electricity consumption during the operation of an electric vehicle, it is important to consider not only the readings recorded by the BEV on-board diagnostic system, but also the nature of the value being analyzed - whether it relates to net consumption (from the traction battery) or gross consumption (from the power grid). In addition, it is crucial to always refer to electric energy consumption in relation to the average speed of travel in a given road cycle, as this has a significant impact on the energy consumption of the vehicle. A separate, important issue is the impact of environmental conditions, in particular the outside temperature, on electric energy consumption. For this reason, these analyses require further continuation and in-depth research on the energy efficiency of modern electric means of transport under various operating conditions.

References

1. International Energy Agency: Global EV Outlook 2025. Paris: IEA (2025). https://www.iea.org/reports/global-ev-outlook-2025
2. Weiss, M., Winbush, T., Newman, A., Helmers, E.: Energy consumption of electric vehicles in Europe. Sustainability **16**, 7529 (2024). https://doi.org/10.3390/su1617752
3. European Commission: Commission Regulation (EU) 2017/1151 supplementing Regulation (EC) No 715/2007 of the European Parliament and of the Council on type-approval of motor vehicles with respect to emissions from light passenger and commercial vehicles (Euro 5 and Euro 6) and on access to vehicle repair and maintenance information (2017)
4. Hooftman, N., Messagie, M., Van Mierlo, J., Coosemans, T.: A review of the European passenger car regulations – real driving emissions vs local air quality. Renew. Sustain. Energy Rev. **86**, 1–21 (2018). https://doi.org/10.1016/j.rser.2018.01.012
5. Engelmann, D., Zimmerli, Y., Czerwinski, J., Bonsack, P.: Real driving emissions in extended driving conditions. Energies **14**, 7310 (2021). https://doi.org/10.3390/en14217310
6. Skobiej, K., Pielecha, J.: Plug-in hybrid ecological category in real driving emissions. Energies **14**(8), 2340 (2021). https://doi.org/10.3390/en14082340
7. Zhao, H., Shi, L., Xu, X., Qiu, J., Li, L., Wang, J., et al.: Impact of shortening real driving emission (RDE) test trips on CO, NO_X, and PN_{10} emissions from different vehicles. Sustainability **16**(21), 9453 (2024). https://doi.org/10.3390/su16219453
8. Hebala, A., Abdelkader, M.I., Ibrahim, R.A.: Comparative analysis of energy consumption and performance metrics in fuel cell, battery, and hybrid electric vehicles under varying wind and road conditions. Technologies **13**(4), 150 (2025). https://doi.org/10.3390/technologies13040150
9. ClearWatt: what is the real-world range for electric vehicles? Battery health, efficiency, and data insights. ClearWatt, May 2025. https://clearwatt.co.uk/blog/what-is-the-real-world-range-for-electric-vehicles-battery-health-efficiency-and-data-insights

10. Berjoza, D., Jurgena, I., Millers, R.: Effect of electric vehicle mass change on energy consumption and the range. Agron. Res. **22**(3), 1089–1097 (2024). https://doi.org/10.15159/AR.24.052
11. Mamala, J., Graba, M., Mitrovic, J., Prażnowski, K., Stasiak, P.: Analysis of speed limit and energy consumption in electric vehicles. Combust. Engines **195**(4), 83–89 (2023). https://doi.org/10.19206/CE-169370
12. Al-Wreikat, Y., Serrano, C., Sodré, J.R.: Driving behaviour and trip condition effects on the energy consumption of an electric vehicle under real-world driving. Appl. Energy **297**, 117096 (2021). https://doi.org/10.1016/j.apenergy.2021.117096
13. Al-Wreikat, Y., Serrano, C., Sodre, J.R.: Effects of ambient temperature and trip characteristics on the energy consumption of an electric vehicle. Energy **238**, 122028 (2021). https://doi.org/10.1016/j.energy.2021.122028
14. Armenta-Déu, C., Cattin, E.: Real driving range in electric vehicles: influence on fuel consumption and carbon emissions. World Electr. Veh. J. **12**(4), 166 (2021). https://doi.org/10.3390/wevj12040166
15. Mamala, J., Graba, M., Stasiak, P.: Energy efficiency of the electric drive system of a passenger car under normal operation conditions. In: Rackov, M., Miltenović, A., Banić, M. (eds.) Machine and Industrial Design in Mechanical Engineering. KOD 2024. Mechanisms and Machine Science, vol. 174, pp. [pages if known]. Springer, Cham (2025). https://doi.org/10.1007/978-3-031-80512-7_90
16. Pielecha, I., Pielecha, J.: Simulation analysis of electric vehicles energy consumption in driving tests. Eksploat. Niezawodn.–Maint. Reliab. **22**(1), 130–137 (2020). https://doi.org/10.17531/ein.2020.1.15
17. Steinstraeter, M., Heinrich, T., Lienkamp, M.: Effect of low temperature on electric vehicle range. World Electr. Veh. J. **12**(3), 115 (2021). https://doi.org/10.3390/wevj12030115
18. Yuan, X., Zhang, C., Hong, G., Huang, X., Li, L.: Method for evaluating the real-world driving energy consumptions of electric vehicles. Energy **141**, 1955–1968 (2017). https://doi.org/10.1016/j.energy.2017.11.134
19. Chłopek, M.: Study on energy consumption by an electric vehicle. Sci. Notebooks UTP, Automation and Diagnostics of Machines and Production Processes – 2nd Scientific Symposium, Bydgoszcz (2012). https://yadda.icm.edu.pl/baztech/element/bwmeta1.element.baztech-article-BGPK-3625-4021
20. Chłopek, Z., Lasocki, J.: Energy consumption testing of an electric vehicle in urban traffic conditions. Sci. Notebooks of the Motor Transport Institute **97**, 1–8 (2014)
21. Instytut Transportu Samochodowego: Influence of traffic conditions on energy consumption by electric vehicles. BazTech Database (2020). https://yadda.icm.edu.pl/baztech/element/bwmeta1.element.baztech-417f7115-187f-45cc-8f5f-d2b8a0935e4f
22. WysokieNapiecie.pl: The whole truth about electric vehicles in winter. (2022). https://wysokienapiecie.pl/84228-cala-prawda-o-samochodach-elektrycznych-zima/
23. Elektromobilni.pl: EV365 research project – preliminary findings. Narzędzia Elektromobilni (2022). https://narzedzia.elektromobilni.pl/aktualnosci/Projekt-badawczy-EV365-pierwsze-wnioski
24. European Commission: Commission Implementing Regulation (EU) 2021/392 of 4 March 2021 on the monitoring and reporting of data relating to CO_2 emissions, fuel consumption and electric energy consumption from passenger cars and light commercial vehicles. Off. J. Eur. Union **L77**, 8–25 (2021)
25. ADAC: Elektroauto Ladeverluste: AC oder DC – was ist effizienter? Allgemeiner Deutscher Automobil Club e.V., München (2025). https://www.adac.de/rund-ums-fahrzeug/elektromobilitaet/laden/ladeverluste-elektroauto-studie/

26. Kieldsen, A., Thingvad, A., Martinenas, S., Sørensen, T.M.: Efficiency test method for electric vehicle chargers. In: Proc. EVS29 - International Battery, Hybrid and Fuel Cell Electric Vehicle Symposium, Montréal, Canada, June 19–22 (2016)
27. Kostopoulos, E.D., Spyropoulos, G.C., Kaldellis, J.K.: Real-world study for the optimal charging of electric vehicles. Energy Rep. **6**, 418–426 (2020). https://doi.org/10.1016/j.egyr.2019.12.008
28. Reick, B., Konzept, A., Kaufmann, A., Stetter, R., Engelmann, D.: Influence of charging losses on energy consumption and CO_2 emissions of battery-electric vehicles. Vehicles **3**(4), 736–748 (2021). https://doi.org/10.3390/vehicles3040043
29. Gebisa, A., Gebresenbet, G., Gopal, R., Nallamothu, R.B.: Driving cycles for estimating vehicle emission levels and energy consumption. Future Transp. **1**(3), 615–638 (2021). https://doi.org/10.3390/futuretransp1030033
30. Grano, E., Villani, M., de Carvalho Pinheiro, H., Carello, M.: Are we testing vehicles the right way? Challenges of electrified and connected vehicles for standard drive cycles and on-road testing. World Electr. Veh. J. **16**(2), 94 (2025). https://doi.org/10.3390/wevj16020094
31. Prati, M.V., Costagliola, M.A., Giuzio, R., Corsetti, C., Beatrice, C.: Emissions and energy consumption of a plug-in hybrid passenger car in real driving emission (RDE) test. Transp. Eng. **4**, 100069 (2021). https://doi.org/10.1016/j.treng.2021.100069

Leveraging GAIA-X for Intelligent Mobility and Transportation Systems in Europe

Sebastian Naumann(✉) and Joachim Schade

Institut F. Automation Und Kommunikation E.V., Magdeburg, Germany
{sebastian.naumann,joachim.schade}@ifak.eu

Abstract. The transportation sector is undergoing a paradigm shift driven by digitalization, sustainability goals, and user-centric mobility solutions. GAIA-X, a European initiative for a federated and secure data infrastructure, presents a transformative opportunity to enable secure, interoperable, and sovereign data sharing across mobility stakeholders. This paper explores how GAIA-X principles can be applied to the transportation sector, focusing on intelligent mobility, logistics optimization, and integrated multimodal transportation systems. We present the architectural framework, use cases, challenges, and policy implications of implementing GAIA-X in transport ecosystems. Two examples illustrate the benefits of GAIA-X.

Keywords: data space · GAIA-X · mobility · transportation

1 Introduction

The digital transformation of the transportation sector is critical to achieving Europe's climate, efficiency, and innovation goals [1]. Data sharing across organizations – OEMs, public authorities, logistics providers, mobility-as-a-service (MaaS) platforms – is essential for optimizing traffic flows, reducing emissions, and providing user-friendly services. However, current data practices are often fragmented, proprietary, and insecure. GAIA-X offers a solution through its federated architecture enabling trusted data spaces, transparency, and interoperability while preserving data sovereignty.

Furthermore, as cities grow and the complexity of urban transportation increases, the need for adaptive and responsive mobility systems becomes more urgent [2]. Traditional models of isolated data control do not support the cross-sectoral integration needed to manage congestion, emissions, and multimodal networks. Therefore, initiatives such as GAIA-X are crucial in redefining the digital landscape of mobility. The seamless integration of smart infrastructure, autonomous systems, electric vehicles, and multimodal logistics can only be achieved through trusted and standardized data sharing platforms [3].

The growing prominence of climate action, urbanization, and digital innovation means that transport systems must evolve to become not only smarter, but also more resilient and inclusive. GAIA-X is positioned as a cornerstone in Europe's ambition to

E. Macioszek et al. (Eds.): TSTP 2025, LNNS 1769, pp. 154–169, 2026.
https://doi.org/10.1007/978-3-032-14094-4_11

lead globally in ethical, efficient, and sovereign digital ecosystems. Its implementation in the transportation sector could address challenges related to sustainability, affordability, safety, and accessibility in mobility services.

2 Transportation Sector Challenges

As the demand for integrated, intelligent, and user-centric mobility solutions grows, the transportation sector faces significant barriers to unlocking the full potential of data. Despite the proliferation of digital platforms and connected services, the current data ecosystem remains fragmented, proprietary, and difficult to navigate. This section outlines key challenges – ranging from data silos and lack of interoperability to trust issues and regulatory complexity – that hinder innovation, scalability, and cross-border collaboration. Addressing these challenges is essential for realizing the vision of a unified European mobility data space powered by frameworks like GAIA-X.

2.1 Fragmented Data Landscape

Data silos across OEMs, logistics operators, municipalities, and MaaS platforms hinder integrated mobility solutions. For example, a lack of access to real-time location data from bus operators or logistics fleets prevents dynamic route optimization or last-mile delivery coordination. Fragmentation also limits the ability to develop regional or cross-border services, making it harder for travelers to access seamless mobility across Europe.

Furthermore, lack of metadata documentation, data quality assurance, and open standards exacerbates these challenges. Even when data is technically available, lack of context or quality limits its usability. There is a pressing need to address the 'dark data' problem – untapped data resources that are underutilized due to fragmentation and lack of discoverability.

2.2 Lack of Interoperability

Proprietary formats and standards prevent seamless data integration and reuse across platforms. The absence of harmonized APIs and metadata descriptions hampers developers and data scientists from building scalable applications. Moreover, interoperability is a prerequisite for creating AI-driven systems that can learn from multimodal data sources, such as weather, road conditions, user behavior, and vehicle telemetry.

Semantic mismatches, inconsistent ontologies, and diverging data models create integration overheads and technical debt. Cross-domain services such as MaaS aggregators and digital twins for cities cannot thrive without strong semantic interoperability.

2.3 Trust and Data Sovereignty

Stakeholders are reluctant to share sensitive operational or user data due to lack of control and unclear governance structures. Without robust mechanisms for consent management and contractual compliance, data holders face legal and reputational risks. There is also concern over data being exploited by third parties without transparency or benefit-sharing mechanisms, which limits innovation and cooperation.

Data sovereignty is especially vital in transportation, where privacy-sensitive and mission-critical data – such as vehicle telemetry, passenger flows, or route maps – must be protected from misuse.

2.4 Regulatory Complexity Compliance with GDPR

ITS Directive, and national regulations add complexity to cross-border data exchange. Different interpretations of data protection and transportation safety laws lead to hesitancy and slow innovation. Regulatory fragmentation also increases costs for startups and SMEs trying to scale mobility solutions across different markets. More coordinated regulatory sandboxes and harmonized certification mechanisms across the EU are needed.

3 GAIA-X Key Principles and Benefits

GAIA-X is a European initiative launched in 2020 to create a federated, secure, and transparent digital infrastructure [4] which tries to overcome the mentioned challenges not only in the transportation sector. Key principles include data sovereignty, interoperability, openness, and trust. The architecture comprises Federated Services (identity, compliance, data catalogues), Data Ecosystems, and the GAIA-X Federation Services (GXFS) stack. For the transportation domain, GAIA-X Mobility Dataspace is being developed under initiatives like Catena-X and Mobility Data Space (MDS).

At its core, GAIA-X promotes a model where data is not centralized in a few dominant platforms, but rather governed by shared standards and principles [5]. Federation of infrastructure providers and data owners allows smaller entities, public administrations, and local innovators to participate equally in the digital economy [6]. This is particularly important in transportation, where public and private actors must collaborate effectively [7].

GAIA-X is not a data platform itself, but an enabler of federated ecosystems. It provides reference architectures, compliance tools, certification frameworks, and semantic interoperability layers. These features empower data spaces across sectors like health, finance, manufacturing, and transportation to achieve independent yet compatible operations under shared governance principles [8]. The following sections explain some of the important basic principles and benefits in more detail.

3.1 Trust and Data Sovereignty

How do we determine who is allowed to use the data? And how do we ensure that only those people actually use the data? GAIA-X allows data providers to define usage constraints using standardized policies and smart contracts, helping to build a foundation of digital trust. Furthermore, GAIA-X could help reduce compliance burdens by offering built-in traceability, auditability, and policy enforcement within data spaces. In the GAIA-X context, trust is not only socially or contractually secured, but also technically and structurally [9, 10]. We will explain this step by step using the GAIA-X principles and tools.

Trusted Identities. Every participant (company, service, API) has a verified digital identity – technically secured via:

- Self-descriptions (describe who you are, what you do, what credentials you have)
- Verifiable Credentials (e.g., Chamber of Commerce and Industry member, GDPR-compliant, ISO-certified)
- Certified Services (e.g., "GAIA-X Trust Anchor")

So, we know who our counterpart is without having to know them personally – machine-readable and verifiable.

Usage Policies: Who is allowed to do what, how, where, and for what purpose? When we provide data, we define technical rules (policies), e.g.:

```
{
    "allowedPurpose": ["Logistics Planning"],
    "geographicRestriction": "EU",
    "retentionPeriod": "24h",
    "commercialUse": false
}
```

These rules are published in a self-description or via the IDS connector – and are machine-readable, interoperable and technically verifiable.

Policy Enforcement: Ensuring that Rules Are Adhered To. This is where the technical trust anchor comes into play:

- GAIA-X/IDS Connectors: Participants use so-called data connectors (e.g., from IDSA, Eclipse Dataspace Components). These connectors only exchange data if both sides are trustworthy and the policies are met.
- Technical Control: If, for example, a user wants to store data outside of Europe, the connector terminates the connection. If a service attempts to store data for longer than 24 h, it will be blocked – or only pseudonymized data will be provided. This isn't wishful thinking – the technology already exists (e.g., IDS Policy Enforcement Points).
- Logging & Audit: Every access can be technically logged. You can prove: Who accessed which data, when, and under which conditions?

For example, someone provides telematics data from his vehicles – but only to certified partners within Germany, and not for commercial reuse. All partners must have a GAIA-X-compliant identity. The request goes through a certified connector that checks the usage policies. The data is only passed through if everything is correct. Access is automatically documented. As result, a customized contract or manual review is not needed; instead, trust is technically implemented in a data-driven, secure, and scalable way.

3.2 Driving Innovation

The next question we want to explore in more detail is to what extent does all this increase innovation capacity. GAIA-X increases innovation capacity because it removes many of the previous barriers to innovation regarding data access, trust, and scalability – especially in data-driven business models. GAIA-X creates a kind of "innovation infrastructure" for the data economy – similar to what the internet once did for communication. It will not be the innovation itself, but it systematically removes the hurdles that have previously blocked innovation [11]. Here are the key points explained in concrete terms:

Easier Access to Data as a Raw Material for Innovation. Previously, data access was usually exclusive, expensive, or legally unclear. Startups and SMEs had little access to relevant data sources. With GAIA-X, data is described and findable in a standardized manner (via federated catalogs). Machine-readable conditions (policies) make access plannable and automated. Even smaller providers can quickly access quality-assured data – e.g., for optimizing traffic, energy consumption, supply chains, etc. More ideas can be realistically implemented because data is no longer "locked up".

Interoperability Reduces Technology Barriers. Previously, every new data partner meant new interfaces, new contracts, and new risks. With GAIA-X, open, standardized interfaces, reusable data models and connectors, clear certification and security frameworks enable building new solutions faster and helps to combine different data (e.g., sensor data with weather data or traffic flows) easily.

Trust Technology Allows Risk-Free Experimentation. Previously, fear of data misuse, GDPR violations, or loss of control blocked many ideas. With GAIA-X, usage policies and technical policy enforcement mechanisms secure intellectual property and legal framework. Stakeholders can "release" services without losing control. Innovations can be made accessible on a trial basis – without long-term risks.

Collaborations are Easier and Faster. Before, complex bilateral contracts and building trust took months. With GAIA-X, partners can quickly assess each other based on self-descriptions and verified credentials. Shared data spaces promote ecosystem innovations, not just individual projects. For example, in a regional mobility data space, the city, logistics, and an AI startup are jointly developing a new delivery forecasting app – quickly, federated, and legally compliant.

3.3 Driving Data Business

Let us explain how business models can be better implemented with GAIA-X. GAIA-X finally brings what is normal for cloud platforms (e.g., pay-per-use, self-service, plug-and-play) to the data-based economy – but with trust, transparency, and sovereignty [12]. With GAIA-X, business models can be implemented better, faster, and more trustworthy because the ecosystem eliminates many of the previous obstacles to data exchange – technical, legal, and economic [13, 14].

So, new data-based business models become possible in the first place. For example, a fleet operator can sell anonymized real-time data from its vehicles – but only for specific purposes (e.g., traffic flow analysis), in limited regions, and for a limited time. With GAIA-X Usage Policies, precisely this can be controlled in a machine-readable manner – creating data-based value-added services without loss of control.

Furthermore, business relationships become more modular and can be established more quickly. Today, we often see individual contracts, lengthy coordination processes and proprietary interfaces. With GAIA-X Self-descriptions, we have machine-readable offer descriptions, verified identities, standardized rights, interfaces, and security. The time to market is drastically reduced; partners can connect on demand.

Instead of submitting to a dominant platform (e.g., Amazon, Uber Freight), companies can collaborate in federated data spaces while retaining data sovereignty. For example, five medium-sized logistics companies, together with cities, are forming a "mobility data space." Each company can monetize its own services without the need for a central platform operator.

Many customers (especially in industry or the public sector) demand GDPR-compliant data usage, proof of origin and usage and proof of sustainability (e.g., CO_2 footprint verification). With GAIA-X-compliant data architecture, this can be provided in an automated and verifiable manner – which is often a prerequisite for tenders or partnerships.

If, for example, a company develops an algorithm to optimize supply chain CO_2 values. With GAIA-X, this service can be used across multiple data spaces (logistics, retail, cities). The business model becomes platform-independently scalable.

4 GAIA-X for Transportation: Technical Framework

GAIA-X enables a federated data infrastructure where stakeholders can publish, access, and use data while maintaining full control. Participants adhere to common rules, trust frameworks, and technical standards. This model allows data providers to define access rules and usage conditions while remaining compliant with European values. Trust anchors and verifiable credentials enable traceability and transparency [15].

A federated data space ensures that no single actor dominates the data value chain. Public authorities retain oversight capabilities while enabling innovation through controlled openness. This paradigm allows public-private collaboration without compromising autonomy or trust [16].

4.1 Architecture

The Eclipse Dataspace Components (EDC) represents a reference implementation for building a Gaia-X-compliant data and services ecosystem [17]. The project, led by the Eclipse Foundation, is available as open source software and is continuously being developed further. Providers who wish to provide corresponding data offerings via a data space created on the basis of the EDC must first create a so-called asset for each data offering [18]. An asset contains both the metadata regarding the corresponding data offering to be shared, as well as information about the original data source and the data transfer protocol. It is important to note that the selection of a data transfer protocol does not refer to the data transfer between provider and consumer, but rather describes the connection of the data source to an EDC connector [19].

All assets are summarized in a federated catalog. Potential data recipients have the option of querying a provider's federated catalog to obtain an overview of the data offered. While assets are used to describe a provider's data offerings, policies are used to regulate access to this data [20]. A policy can specify in fine-grained detail which consumers are allowed to access certain data offerings (assets) and under which conditions. Examples include time-controlled access, setting limits for a limited number of data downloads, or restricting access for consumers from certain geographical areas, such as the European Union.

Finally, digital contracts – so-called contract definitions – are defined, which establish the assignment of data offerings (assets) to access rules (policies). These are applied during the automated digital contract negotiation process between a requesting consumer and the provider.

The EDC connector is used to exchange data in a data space set up using the EDC. Each participant in the data space who wishes to provide or use data via the data space must do so via an EDC connector, which they either implement themselves or provide as a service from third parties (see Fig. 1). Regardless of the data transfer protocol used for the original data source, the basic process for exchanging data with the EDC connector always follows the same pattern.

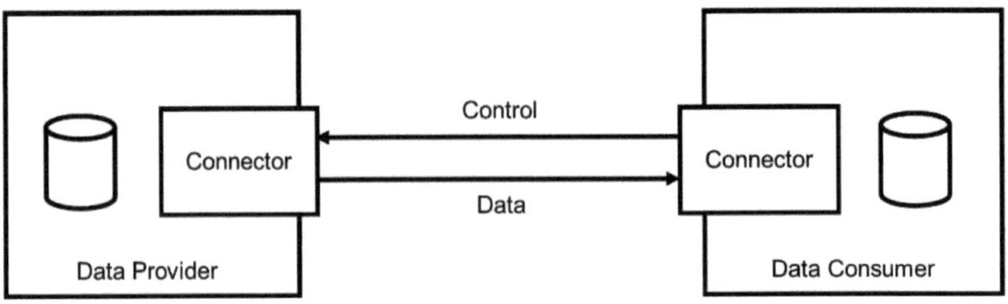

Fig. 1. Connectors in GAIA-X

According to Fig. 1, a technical data space of two entities: a data consumer and a data provider. Both, provider and consumer are decentralised. They can be separated computers like a home laptop as well as a server hosted at a professional hoster. Figure 2 provides the sequence of interaction between a data provider and a data consumer when transferring data. Sovereignty is guaranteed: no one else can access a provider's data if he does not want so. Through protocols and other mechanisms, the provider determines, for what his data is used by whom.

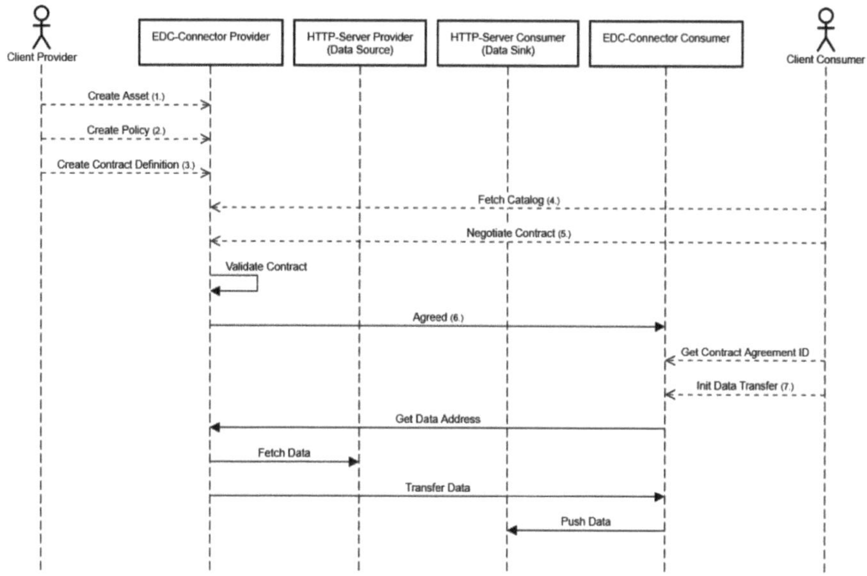

Fig. 2. Data transfer using the EDC connector

This modular stack supports composable architectures, allowing plug-and-play integration of services. By separating concerns – identity, policy, compliance, semantics – it provides resilience and scalability.

4.2 Integration with Existing Systems

GAIA-X builds on existing standards like DATEX II, NeTEx, and Open Mobility Data standards, allowing backward compatibility. Existing infrastructures such as traffic management centers and fleet management platforms can be enhanced with GAIA-X connectors, enabling gradual migration rather than full replacement. This modular approach reduces risk and encourages experimentation.

Pilot projects across European cities can serve as learning grounds for developing best practices in integration. The success of integration hinges on strong stakeholder dialogue, clear benefit articulation, and minimal disruption to legacy systems.

As a newbie, there are several concrete ways to find relevant data spaces and data sources that fit your business model. First, you can search federated catalogues like Mobility Data Space (MDS) [21] for Mobility & Logistics and Catena-X [22] for Automotive & Supply Chains. You can register there, filter data offerings, and often request a connection directly. Second, there are national GAIA-X (e.g. (GAIA-X Hub Germany) hubs but also topic-specific groups where you can get advice on suitable data spaces, contacts to other organisations and use case partners as well as information on funding projects or test fields. Finally, there are a lot of events and communities, e.g. GAIA-X Summit, IDSA Summit (International Data Spaces), FIWARE Global Summit and Sector-specific GAIA-X Meetups (Mobility, Energy, Industry) where you can connect with potential partners and get to know real use cases.

5 Examples

GAIA-X enables a lot of uses cases and applications in the transportation sector, e.g. [23]. The following two examples (a) Collaborative Freight ETA Optimization and (b) Short-term Warning Service are intended to illustrate the application of GAIA-X including their concrete benefits.

5.1 Collaborative Freight ETA Optimization

Multiple logistics companies agree to share anonymized real-time truck location and freight status data in a GAIA-X-compliant logistics data space. The goal is to optimize estimated times of arrival (ETAs) and dock scheduling at logistics hubs (e.g., warehouses, ports, cross-docking centers). Participants are:

- Logistics Providers A, B, C (data providers)
- Warehouse Operator X (data consumer and co-provider)
- AI-based ETA Service (data service provider)
- GAIA-X Data Space (e.g., part of Mobility Data Space or Catena-X)

By GAIA-X trust and control, each truck's data is shared only under defined usage policies, such as:

- Purpose: "ETA prediction and slot scheduling only"
- Geo-restriction: "EU servers only"
- Data retention: "Max. 48h"
- Commercial use: "No resale of raw data"

Access is enforced via GAIA-X-compatible Data Connectors (e.g. Eclipse Dataspace Components):

```
Example API: GET /truck-status
OpenAPI-style Example (simplified)

GET /truck-status

Description:
Returns current truck location and freight status for au-
thorized logistics partners.

Query Parameters:
  truckId: string (optional) - filter by known truck ID
  carrier: string (optional) - filter by logistics pro-
vider
  updatedSince: datetime (optional) - only recent updates

Headers:
  Authorization: Bearer <GAIA-X-compliant token>
  X-Usage-Purpose: "ETAOptimization"
```

This API lets authorized logistics partners or services retrieve real-time information about trucks such as current location (latitude/longitude), freight status (e.g., "in-transit", "delayed", "delivered"), estimated time of arrival (ETA) and carrier information (e.g., the transport company).

The HTTP GET Request means the client (e.g. a warehouse IT system) asks for data without modifying anything. We can use filters (Query Parameters) to narrow the request: `truckId` for getting data for a specific truck, `carrier` for seeing trucks from a certain company or `updatedSince` for seeing updates newer than a given timestamp (for polling).

The headers carry access control and purpose information as `Authorization` – this uses a GAIA-X-compliant token, like a verifiable credential, to ensure only authorized users access the data – and `X-Usage-Purpose` – this declares why the client is using the data. This is checked against the provider's usage policy (e.g., only allowed for "ETA optimization").

We get back a json object like this:


```
{
  "data": [
    {
      "truckId": "LKW-9834",
      "location": {
        "lat": 51.2321,
        "lon": 6.7928,
        "timestamp": "2025-05-19T12:23:00Z"
      },
      "freightStatus": "in-transit",
      "estimatedArrival": "2025-05-19T14:30:00Z",
      "carrier": "FastTrans GmbH"
    }
  ],
  "meta": {
    "policyCompliant": true,
    "dataProvider": "GAIA-X ID: fasttrans-gmbh.de"
  }
}
```


This tells us, that (a) truck LKW-9834 is currently driving somewhere in Germany, (b) it is expected to arrive at 2:30 PM, (c) the data is compliant with GAIA-X policies (e.g., usage restrictions, identity) and finally (d) it came from a trusted logistics provider, verifiable via their GAIA-X ID.

Table 1. GAIA-X value creation

Without GAIA-X	With GAIA-X
Data silos, no real-time collaboration	Standardized, secure data sharing
Manual ETA estimates, inefficiencies	AI-based dynamic ETA across all participants
Lack of trust in shared data	GAIA-X Self-Descriptions & Policies ensure authenticity and rules
High onboarding costs per partner	Federated, reusable trust framework and connectors

Table 1 lists the values created by using GAIA-X. Without GAIA-X data might be shared in insecure ways, we do not know if it's real, current, or allowed to be used and each partner integration is manual and custom. With GAIA-X connectors control access and enforce who gets what data for what purpose, trust is built-in via digital identities and usage policies and partners can plug in and share data confidently, even across competitors.

5.2 Short-Term Warning Service

A second example of GAIA-X benefit is the provision of short-term traffic disruptions in the form of warning messages. These warning messages represent an important complement to longer-term traffic disruptions, such as roadworks, and are of interest for a variety of purposes. For this purpose, the ZEBRA construction site information management tool developed by our own was upgraded and expanded to include functions for entering short-term warning messages (Fig. 3).

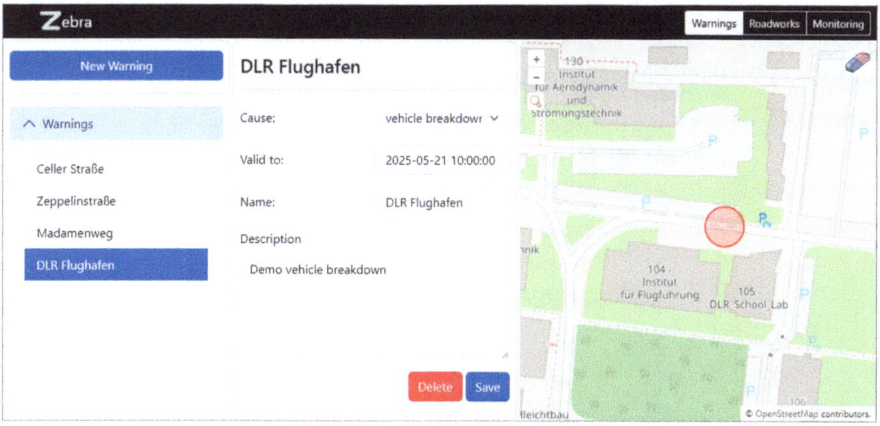

Fig. 3. Warning registered in the ZEBRA tool

In this real example, only a part of the whole data flow is implemented GAIA-X compliant, as Fig. 4 shows.

The warnings entered are sent out via mobile network and ETSI ITS-G5 (V2X) as DENM messages (Decentralized Environmental Notification Messages). They are used for **ODD** (Operational Design Domain) m**atching and ODD compliant routing in the context of** automated driving, where areas are defined within an autonomous vehicle is permitted to move, as well as presenting the warning information to the driver (see Fig. 5).

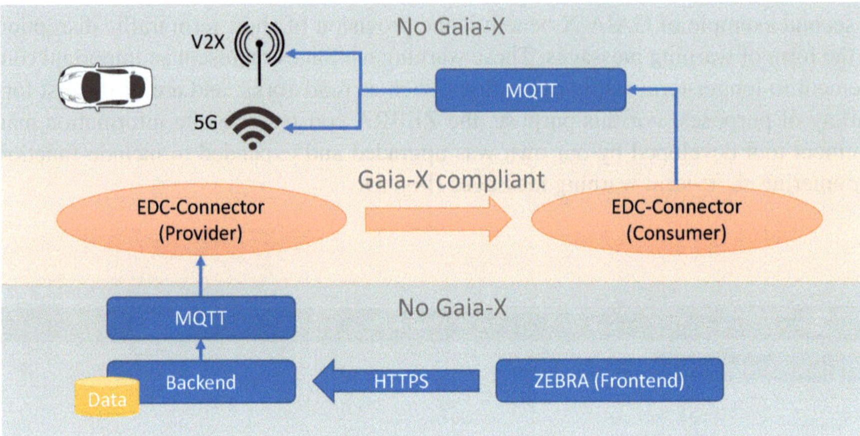

Fig. 4. Data flow

Fig. 5. Use of warning messages in the vehicle

6 Conclusions

Implementing GAIA-X in transportation could significantly impact employment, innovation, public service delivery, and competitiveness. SMEs and startups can participate more easily in federated ecosystems, lowering market entry barriers. Public agencies can deliver more efficient, citizen-centered mobility services. Enhanced data access and transparency also support scientific research, public accountability, and democratic participation in infrastructure planning.

The democratization of data through GAIA-X principles may also redistribute value creation in the transport data economy, reducing dependence on non-European cloud and platform providers. This could help retain value chains and intellectual capital within Europe. Adoption of GAIA-X in transport depends on multi-level stakeholder cooperation. This includes public authorities, transport operators, infrastructure providers, data intermediaries, users, and tech developers. Participatory governance models must ensure representation of diverse interests and transparent rule-making. Regional mobility alliances and public-private innovation labs could serve as incubators for GAIA-X mobility applications. Cross-border testbeds – especially in border regions – can explore interoperability and regulatory harmonization.

Despite its promise, GAIA-X faces challenges including low awareness among mobility actors, complexity in aligning semantic models across countries and domains as well as risk aversion among public authorities and SMEs. In particular, the involvement of non-European cloud providers such as Google, Microsoft, and Amazon, as well as the bureaucracy associated with the project, are the focus of criticism. Scaleway CEO Yann Lechelle justified his withdrawal from the Gaia-X project with the obstruction of major US IT companies. They would block and sabotage any progress towards a manufacturer-neutral, portable model through delay [24]. Frank Karlitschek declared on the occasion of Nextcloud's withdrawal in February 2025 that "the original goal of establishing a European cloud alternative to the American hyperscalers is no longer being discussed today" [25].

Despite these critical voices, the fundamental advantages of GAIA-X remain undiminished and point to the future direction of intelligent transport systems. The transformative potential for Europe's transportation sector is undisputed. By enabling secure, sovereign, and interoperable data ecosystems, the way for a smarter, greener, and more inclusive mobility future is paved. Success will depend on collective commitment, standardization, and the will to innovate collaboratively.

Funded by the European Union
NextGenerationEU

Supported by:
Federal Ministry for Economic Affairs and Climate Action
on the basis of a decision by the German Bundestag

Acknowledgements. The work presented in this paper has been carried out within the German national research project GAIA-X4AMS funded by the German Federal Ministry of Economic Affairs and Climate Action under the grant number 19S21004S. The authors would like to thank for the financial support.

References

1. European Commission: Sustainable and Smart Mobility Strategy (2021). https://transport.ec.europa.eu/transport-themes/mobility-strategy_en
2. OECD: OECD Urban Studies – Smart City Data Governance (2022). https://www.oecd.org/content/dam/oecd/en/publications/reports/2023/10/smart-city-data-governance_fc19e878/e57ce301-en.pdf
3. European Data Strategy: European Commission (2020). https://commission.europa.eu/strategy-and-policy/priorities-2019-2024/europe-fit-digital-age/european-data-strategy_en
4. Tardieu, H.: Role of Gaia-X in the European data space ecosystem. In: Designing Data Spaces: The Ecosystem Approach to Competitive Advantage, pp. 41–59. Springer, Cham (2022). https://doi.org/10.1007/978-3-030-93975-5_4
5. Scerri, S., Tuikka, T., de Vallejo, I.L., Curry, E.: Common European data spaces: challenges and opportunities. Data Spaces: Design, Deployment and Future Directions, pp. 337–357 (2022)
6. Otto, B., ten Hompel, M., Wrobel, S. (eds.): Designing Data Spaces – The Ecosystem Approach to Competitive Advantage. Springer, Cham (2022). ISBN 978-3-030-93974-8. ISBN 978-3-030-93975-5 (eBook). https://doi.org/10.1007/978-3-030-93975-5
7. Schlueter Langdon, C., Schweichhart, K.: Data Spaces: First Applications in Mobility and Industry. In: Otto, B., ten Hompel, M., Wrobel, S. (eds.) Designing Data Spaces, pp. 493–511. Springer, Cham (2022). https://doi.org/10.1007/978-3-030-93975-5_30
8. Curry, E., Scerri, S., Tuikka, T.: Data Spaces: Design, Deployment and Future Directions, p. 357. Springer, Cham (2022). https://doi.org/10.1007/978-3-030-98636-0_1. https://library.oapen.org/bitstream/handle/20.500.12657/58395/1/978-3-030-98636-0.pdf
9. Fraunhofer ISST: Digital Sovereignty (2022). https://www.isst.fraunhofer.de/content/dam/isst/publikationen/Industrial-Data-Space/Fraunhofer-IDS_Digitale-Souver%C3%A4nit%C3%A4t_Brosch%C3%BCre_DE_2016.pdf
10. BMWK: Sovereign Data Infrastructures for Mobility (2023). https://www.bmwk.de/Redaktion/EN/Dossier/gaia-x.html. The Gaia-X Ecosystem - A Sovereign Data Infrastructure for Europe (2023). https://www.bmwk.de/Redaktion/EN/Artikel/Digital-World/GAIA-X-Use-Cases/mobility-data-interoperability-and-data-sovereignty.html
11. Adler-Nissen, R., Eggeling, K.A.: The discursive struggle for digital sovereignty: security, economy, rights and the cloud project Gaia-X. JCMS J. Common Mark. Stud. **62**(4), 993–1011 (2024)
12. The Role of Data Spaces in the Digital Economy. White Paper. https://gaia-x.eu/wp-content/uploads/2025/03/White-Paper_The-Role-of-Data-Spaces-in-the-Digital-Economy-1.pdf
13. Drees, H., Kubitza, D.O., Lipp, J., Pretzsch, S., Schlueter Langdon, C.: Mobility data space – first implementation and business opportunities. ITS World Congress, October 2021
14. Sorescu, A.: Data-driven business model innovation. J. Prod. Innov. Manag. **34**(5), 691–696 (2017)
15. ISO/TC 204 & CEN/TC 278 Standards. https://www.itsstandards.eu/aboutus/
16. Gaia-X AISBL (Editor): Gaia-X Architecture Document (2022). https://www.gaiax.es/sites/default/files/2022-01/Gaia-X_Architecture_Document_2112.pdf
17. Heinbach, C., Gösling, H., Thomas, O.: The Hitchhiker's guide to urban spaces–conceptualizing a Gaia-X-enabled co-bility hub combining public transport, crowd mobility, and last mile logistics. In: Designing Futures: Zukünfte gestalten. Gesellschaft für Informatik, INFORMATIK 2023, pp. 1359–1371 (2023)
18. Dam, T., Klausner, L. D., Neumaier, S., Priebe, T.: A survey of dataspace connector implementations. arXiv preprint arXiv:2309.11282 (2023)

19. Siska, V., Karagiannis, V., Drobics, M.: Building a dataspace: technical overview. Gaia-X Hub Austria (2023)
20. Pampus, J., Jahnke, B. F., Quensel, R.: Evolving data space technologies: lessons learned from an IDS connector reference implementation. In: International Symposium on Leveraging Applications of Formal Methods, pp. 366–381. Springer, Cham (2022). https://doi.org/10.1007/978-3-031-19762-8_27
21. Mobility Data Space: Technical Whitepaper (2024). https://www.mobility-data-space.de/
22. Catena-X Consortium: Data Ecosystem Architecture (2023). https://catena-x.net/
23. Kremer, M., et al.: An intelligent arrival time prediction service in a federated data ecosystem: the minimum viable demonstrator of the GAIA-X 4 ROMS research project. Available at SSRN 4331859 (2022)
24. Mahn, J., Wölbert, C.: I don't expect Gaia-X to deliver what we need (orig. Ich erwarte nicht, dass Gaia-X liefert, was wir brauchen.) In: c't – Magazin für Computertechnik. Nr. 1, 18. Dezember 2021, S. 14 ff (2021)
25. Kroker, M.: Nextcloud-Chef Frank Karlitschek: Gaia-X is dead (orig. Gaia-X ist tot). WirtschaftsWoche, 7.2.2025 (2025)

Digital Twin of a Specialized Off-Road Vehicle for Peoples with Special Needs

Marcin Staniek and Ireneusz Celiński(✉)

Faculty Transport and Aviation Engineering, Silesian University of Technology, Gliwice, Poland
{mstaniek,icelinski}@polsl.pl

Abstract. The article presents proposals for the design and solution of a digital twin of a specialist off-road vehicle, intended for people with special needs. Unlike typical road vehicles, off-road vehicles move in an environment in which vehicle-road interactions are different, much more dynamic and variable along the entire length of the displacement. These characteristics are especially important in the context of users with special needs, who may require special supplies and protection, according to individual needs. The safety of off-road vehicle operation is at the forefront here. On the other hand, traffic on the mountain trail is slow and generally sparse. The proposed solution of a digital twin of a specialist off-road vehicle, presented in this article, addresses this challenge. In the article, we discuss a vehicle-road interaction model and other characteristics using data from a set of appropriate sensors analyzing the vehicle's movement on a specific route in real time. This approach allows, based on the data obtained from the digital twin, the optimization of vehicle design solutions and the traffic control system. It also allows the design of solutions for devices supporting safe and reliable operation of an off-road vehicle for people with special needs.

Keywords: Digital twins · off-road vehicles · people with special needs

1 Introduction

Digital twin (DT) mapping of functional structures is now almost universal and common. It applies not only to science but also to industry in its broadest sense. According to press reports, the DT-related market will grow by nearly $40 billion by the end of 2025. Is this market unique to transportation systems? In this case, it is important to note the need to map static infrastructure elements and dynamic processes. Static mappings concern infrastructure that changes very slowly. Dynamic mappings of the traffic burdening this infrastructure are complex and sometimes difficult to implement. Unlike other systems, knowledge about traffic engineering theory is constantly updated and often controversial. Road traffic is described as a so-called unregulated process. This refers not only to the spatiotemporal characteristics of traffic at the micro level, but also to all scales of mapping, from macro to nano and meso.

This is not a simple process of mapping elements and their movements. Interactions are more important here. Therefore, the digital twin of a road network is not trivial. The

E. Macioszek et al. (Eds.): TSTP 2025, LNNS 1769, pp. 170–193, 2026.
https://doi.org/10.1007/978-3-032-14094-4_12

static nature of infrastructure is also conventional nowadays, and traffic dynamics are constantly increasing. These processes are compounded by the transformation associated with electromobility, the expected spread of autonomous vehicles (AVs), issues related to sustainable development, and a few transport-related problems.

A digital twin is a concept representing real-world objects, constructed for a specific purpose. Such a purpose may be decision-making. Justified decisions regarding the represented real-world object. It is generally accepted that a real-world object is defined as physical, and its digital representation in the form of a digital twin is a so-called virtual object. This does not mean that a digital twin cannot contain physical elements to some extent, and vice versa. A digital twin is a conglomeration of data, the object's model, the object's relationships with the model, the control system, ICT networks, and many others. The physical model of an object is created through documentation, currently reduced to the process of digitization. This is collected in various forms, such as 3D laser scanning. It is supplemented by data on the object's properties and behavior. Sometimes it is said that the digital twin and the physical object are permanently connected, which is debatable. In this system, a unidirectional or bidirectional flow of information is possible. Data exchange makes digital twins what they are. A digital twin is data and their structures, updated over time. Data analysis enables prototype testing outside the actual facility. This separation reduces cost and operational challenges associated with real-world facilities. For this second reason, digital twins are used in the cybersecurity industry, where physical separation between these facilities is utilized. Data monitoring allows for real-time control within the object-model system. In the literature review presented below on digital twins in transportation systems, due to the subject matter of this article, we will focus on presenting research on road infrastructure mapping and selected vehicle models. Digital twins are commonly associated with construction projects. Building information modeling (BIM) enables the management and use of dynamic quantitative, geometric, spatial, and documentary information [1]. BIM supports decision-making in infrastructure management, including road infrastructure. One study on this topic proposed a framework for road infrastructure digital twins. This framework aims to ensure optimal decision-making in road operation and maintenance. The authors of this article discuss how DT supports and integrate information and processes in road management [1]. DT infrastructure can be used to simulate the construction, management, and maintenance of various road assets throughout the LCA [2]. Digital reconstruction techniques use sensors to create a DT of physical infrastructure. Such methods are used to study asphalt pavements [2]. In this case, mapping of the local road network is used. One study on this topic discussed the further development of this type of concept, considering road surface temperatures [2]. Techniques such as LIDAR, unmanned aerial vehicles (UAVs), and artificial intelligence (AI) are used here, allowing for the quantitative determination of road geometry over large areas [2]. On the other hand, a photogrammetric reconstruction technique based on a neural network and an environmental sensor is used [2]. Temperature sensors are used to measure surface texture and temperatures, as well as environmental conditions [2]. Combining environmental and physical data provides essential management data. This data enables informed decisions regarding road maintenance [2]. BIM also encompasses the multi-level of details concept (LOD) for road DT. This clearly organizes the various DT applications in road engineering. The problem stems from various and partially contradictory definitions of LOD. According to the authors of this article, these create misconceptions that hinder faster adaptation of digital methods in infrastructure construction and maintenance [3]. The next article on this topic describes a specific DT concept for roads [4]. In this context, DT has been applied in Smart Cities (SC). The authors of this study propose the deployment of so-called Digital Twin Boxes (DTBs) on roads. Each of these devices consists of a 360° camera and a set of IoT devices connected to a controller. DTB creates a DT as a physical road asset. It operates in real time. Video data, GPS locations, and temperature and humidity measurements are transmitted using cloud technology. This data is used for real-time monitoring. An object detection process is also carried out to extract all indicated objects from the captured stream. A tracking module is used to identify relevant pedestrians and vehicles [4]. The authors of this study conducted experiments in the real world, and in their opinion, the results obtained show the effectiveness of the proposed solution, which is being implemented under AV [4]. Other authors point out that currently, traffic management is increasingly being handled by AI-supported DT instead of control centers [5]. They described examples of implementing such technologies within ITS. This served as the basis for developing a reference model for such a system [5]. DTs are often used in highway management. This requires recording the broadly understood so-called highway condition. As the authors of this study point out, examining highways is expensive, dangerous, and time-consuming [6]. Therefore, they proposed an approach to creating a highway DT using map data. Such a DT consists of the main elements of the highway: horizontal, vertical, cross-section, traffic lanes, etc. The solution was tested on a section of the A1 motorway in the UK. According to the authors, such DT requires minimal human input and is very accurate. According to their data, the average vertical deviation per square meter between the surface of the generated DT and the actual data was at the level of a centimeter [6]. Another article on this topic addresses the recently popular need to improve the durability and resilience of infrastructure. Its authors presented the current scope of DT technology applications in rail and road networks [7]. They also presented conclusions on how these applications can contribute to increasing their durability and resilience. To this end, they propose a systematic review of the literature on digital twins related to rail and road infrastructure networks. The obtained results show that digital twins research in this field is still scarce, and most applications in rail and road networks focus on their operation and maintenance [7]. Digital twins are used in traffic studies. The goal is effective traffic management to improve road safety (RTD). Furthermore, digital twins are used there to improve the capacity and, more broadly, the efficiency of the transport system. The authors of one of the articles in this area discuss the technologies used to create these digital twins, as well as methods and algorithms for data based on which informed decisions can be made. The authors discuss the potential benefits and challenges associated with using digital twins to analyze traffic flows [8]. The authors of another study point to the digitalization, automation, and robotization of road control and maintenance technologies as means to facilitate the implementation of digital twins. In this context, they presented a decision support tool based on digital twins describing the main components and functionalities [9]. Such DT describes actual road resources to integrate various data sources and support data processing. Their decision support tool can analyze the collected information and calculate the condition of the road surface to develop optimal intervention plans [9]. The authors described the application of the proposed architecture to road surface maintenance. Data such as roughness index and lateral force coefficient were used. The data is processed using AI clustering techniques. The result of AI implementation is the division and grouping of road sections according to status and quality index [9]. Other authors point out that driving requires the driver to make constant decisions. However, humans can cope with poor-quality information, while machines require reliable information [10]. According to these authors, DT should contain information ready for delivery to AV systems. In this context, the authors of the study proposed a list of requirements that an information model for AV should meet. A system architecture with three different implementations was proposed and verified, addressing several different decision support applications [10]. DT implemented in road traffic uses a microscopic traffic model. Such DT utilizes publicly available map data and traffic simulations, e.g., those generated in SUMO software. Alternatively, a few commercial applications, e.g., PTV, can be used for this purpose. Traffic study data for DT can be obtained using drones. Validation of this approach has demonstrated the validity of such implementations [11]. The author of this paper demonstrated the usefulness of the traffic model in a case study. In the author's opinion, authorities should be able to use the developed traffic model to evaluate infrastructure, improvements, and prioritize projects based on benefits [11]. Another study in this area describes the definitions, concepts, and applications related to DT implementation in civil infrastructure systems (CIS). This topic is presented in relation to topics such as transportation, energy, telecommunications, water, waste, and SC. The authors conducted a literature review, a series of semi-structured interviews, and a survey. The result of this work is an improved definition of DT and a set of recommendations for both future research in this field and industry practices [12]. According to subsequent authors, DT in transport infrastructure is becoming an increasingly important tool due to its potential to increase the reliability and efficiency of resources. Consequently, DT provides up-to-date information on the condition of road infrastructure and existing hazards. Therefore, DT is the basis for decision-making regarding infrastructure. In this context, the authors of the article presented an innovative perspective on the design, prototyping, and testing of DT for bridges and other road infrastructure. The proposed methodology considers the potential for developing DT for road infrastructure, including stratigraphic analysis, pavement condition monitoring, and bridge structure analysis [13, 14]. Below, we describe the literature that relates to the topic of road vehicles and their users. These three elements, road, vehicle, and user can form separate or integrated DTs. In this context, the issue of DT as a digital representation that allows for the description of LCA of complex elements, systems, and processes is discussed [15]. The authors of this paper argue that this methodology is used in the automotive industry for this reason. DTs are widely used, from design to vehicle production. The role of DTs does not end there. This technology is used in everyday operation, enabling improvements in vehicle performance, especially in terms of safety. The technology is widely used in electric vehicles (EVs) [15]. One such application is the design of a DT system (DTS) using a virtual driving simulation environment. In this system, DTs communicate with each other and with a service system that collects data for the DT. The DT cooperates with a safety controller or electronic control unit (ECU). The purpose of data exchange is to evaluate automotive systems. In this project, the authors create a so-called Hardware-In-the-Loop (HIL) test setup, which enables real-time ECU testing. Data is sourced from virtual sensors located in a virtual simulator to the physical ECU. The project aims to analyze vehicle safety [16]. DTs enable lifecycle extension. In the automotive industry, they are used to improve the design, construction, and operation of EVs. Smart EVs should contribute to emission reduction. This requires the use of innovations in the form of data optimization. The authors of the following article point out that only the development of data analytics and the IoT has accelerated the implementation of data optimization. For example, EV batteries require analysis of their charge and technical status. In this context, the authors of this study address the issues of monitoring, tracking, battery and charging management, communication, and ensuring safety and reliability [16]. In addition, they analyze trends in energy storage technologies in EVs. DT technology enables LCA analysis, increasing the efficiency of battery management [17]. According to other researchers, DT represents a breakthrough on the Internet of Vehicles (IoV) [18]. Technology improves vehicle design and driver assistance systems. To this end, the authors reviewed recent scientific papers on the importance of DT in the field of IoV. The study analyzed four key sectors: industrial production, driver assistance technologies, ITS, and resource management [19]. The aim of the review was to stimulate innovation in the field of IoV by providing relevant knowledge [18]. DT in the automotive industry is being implemented to minimize financial and time losses by streamlining costly and time-consuming EV design and implementation processes [19]. During EV operation, DTs are used to monitor, evaluate, and optimize vehicle performance based on real-time road data. In one study, the authors proposed a Ceryan motorcycle model to compare the performance of different engine types. The comparison focused on energy consumption and acceleration at various inclination angles, using real-time road data. The study used references to the World Motorcycle Test Cycle (WMTC). The PMSM brushless motor proved to be the best performer in terms of performance [19]. The use of DTs in vehicles is enabled by increasingly better automotive buses with cheaper, advanced electronics. In one study, a laboratory bus was built that faithfully mimicked a real-world bus. The network layer utilized a CAN (Controller Area Network). MATLAB-defined models were used for various subsystems, such as the anti-lock braking system (ABS), the drive system, and the electric power steering. In this study, the DT evaluation of the vehicle bus is achieved by providing realistic input data, recording, and repaying the vehicle's network traffic. The authors of the study claim to have obtained a good correlation between DT output data and data collected from a real car [20]. Another review of DT implementations for vehicles covers various prototyping approaches, focusing on intelligent vehicle systems [21]. The authors of another study point out that, however, there is no common understanding of the role of DT in the automotive industry [22]. Furthermore, they point out that there are no common modeling techniques for creating DTs for cars. To bridge the gap between scientific literature and the industrial perspective, the authors conducted a survey among experts from Germany. The aim of the survey was to examine the objectives and possibilities of DT and modeling techniques related to DT engineering and maintenance at various stages of automotive development. The survey results indicate that DTs are considered most useful in the usage and support phase. Simulation models, source code, and business process models are considered most important for DT integration [22]. Automotive diagnostics is another area of DT application. The authors of this article point out that there is a lack of current research addressing this issue in the context of service-oriented vehicle architecture. Therefore, they developed DT for automotive diagnostics. They explained the scenario of its application, describing the challenges associated with integration. They identified areas where general techniques for mitigating integration problems have not yet been developed, and a specific action plan for DT diagnostics [23]. Other authors point out misunderstandings regarding the concept of DT, which is based on physics-based models. The authors considered the application of vehicle system dynamics (VSD) theory as the basis for DT development and implementation. They conducted a review in this area, the results of which show that most of the work focuses on the development of DT at the highest level, and that more comprehensive research and analysis is needed in the field of applying VSD in the design and construction of DT [24]. The authors summarized possible directions for the development of future frameworks in relation to disciplines related to VSD [24]. Another DT application concerns an advanced platform for electric trucks. This consists of universal multi-layer DT architecture and multi-disciplinary software packages. The DT platform can implement all functionalities: physical, communication channel, and data fusion. Within the platform, a pre-DT energy consumption estimation model was developed to evaluate conceptual designs. According to the authors, a holistic DT framework enables industrial-level support for new designs for electric truck applications within the European project called Next_ETRUCK [25]. Another broad area of DT application is motorsports. The goal of implementation here is rapid prototyping and testing of new solutions. This is achieved by imitating professional racing drivers based on data collected on a hardware simulator called Driver-in-the-Loop (DiL) [26]. In this application, the authors proposed reinforcement learning to develop strategies resembling human behavior. The authors of this study fitted a probabilistic model to expert demonstrations. They used the data as context for a reinforcement learning agent with context-specific states and rewards. They call this a behavioral cloning agent (BA). According to the authors, their agent outperforms the professional driver used to collect data by an average of approximately 0.4 s per lap. This is the first case known to the authors in which a BA has outperformed a professional racing driver. The agent indicates an increase in performance, facilitating the creation of DT simulations of DiL [26]. DT applications very often concern two-wheelers. In one such study, an electric bicycle was experimentally tested based on energy consumption on defined urban road profiles. The all-electric mode was selected for the study. The DT model was built based on the drive system model to estimate instantaneous electricity consumption depending on voltage, current, torque, and road type classification [27]. The authors designed and programmed a data logging system to measure electric energy consumption. Road profiles were categorized by road surface type and slope. Energy consumption data were used in drivetrain modeling to create a physical model of an electric bicycle (EB). These data served as input to a neural network model representing the DT model [27]. Subsequent work on this topic aims to address EB range issues. The DT inputs include GPS coordinates and altitude recorded using a mobile device, charging power recorded using a smart plug, and recorded weather conditions. The authors formulated a theoretical model of the bicycle's dynamics and energy exchange. A methodology for fitting unknown parameters was applied using dedicated tests. The DT was validated in the field, and its estimated range was compared with the actual distance traveled during a full discharge. Improvements are discussed, as well as the potential of the DT for predicting range under variable road and weather conditions. This is more accurate than the last-stage predictions used in EB displays [28].

Another area of DT applications is off-road vehicles. In this regard, the authors of the study analyzed two different robot simulation frameworks. The first one is based on Gazebo, an open-source 3D robotics simulator used to test and validate algorithms developed within ROS. The second one uses mechanical vehicle assembly in MSC Adams (multi-body simulation). The models were developed for a tracked robot that uses an innovative articulated passive suspension system. In addition, a Gazebo model was developed for a four-wheel drive robot with a steering system, including robot sensors such as GNSS, IMU, and visual sensors. The authors of this study presented details of the design and implementation of the model. An agricultural scenario was selected as a use case. According to the authors, the Gazebo platform can serve as a suitable robot simulation platform for DT vehicle development, if it considers real-time sensor measurements [29].

Another issue addressed in the field of DT is the efficiency and safety of operators in the construction industry. DT can provide added value to heavy construction equipment operators. In this context, the authors of this study presented a DT architecture implemented in an off-road vehicle, which provides the operator with real-time feedback [30]. According to the authors, monitoring the interactions between the off-road vehicle and employees will ensure a safer and more productive work environment. A methodology based on environmental sensing and vehicle condition monitoring was used. The machine's movement and status are realistically visualized in real time in the Unreal environment. The modular architecture enables communication between sensors and the camera, data processing, and DT visualization [30]. Besides the infrastructure and the vehicles using it, another piece of the puzzle in describing DT for transport systems is the human, as a driver, pedestrian, or operator. Various concepts for creating Human DT (HDT) have been proposed in the literature. In this context, various works describe the basic assumptions, the HDT's design, its characteristics, and the methods of HDT construction. These include creating a human model template, its unification, individualization, combining data with data-driven models and integration [31]. One such study presented two projects in which HDT was used to solve practical problems [31]. Human Digital Twins (HDT) are a rapidly developing technology with significant potential in various fields, not just transportation. HDTs expand the traditional understanding of DT by presenting humans as the primary physical entity. The authors of this study note the ambiguity of the HDT definition and the lack of guidelines for their design [32]. Therefore, they summarized the latest developments in the HDT field to help future DT developers by proposing the first cross-domain definition of HDT based on their characteristics and eleven key design issues [32]. In this context, in construction, there is a lack of knowledge on how to implement human-system interactions. According to the authors of another study, this limits the potential of HDT to support decision-making. The authors of this study consider ways to use construction control centers as DT interfaces. The goal of this approach is to increase the accuracy and efficiency of decision-making in the DT construction process. The authors developed a framework for human-DT interfaces and presented their evaluation in a real-world construction project [33]. The authors argue that control centers can act as dynamic interfaces in the DT ecosystem, improving decision-making efficiency. This emphasizes the role of DT interfaces in decision-making support [33]. The next article addresses human interaction with intelligent machines. According to the authors of this study, HDT enables technology designers to model human interactions with intelligent technical artifacts [34]. The authors constructed a human DT for Minsky's M-Machine. The abstract conceptual structure of this machine and its generality allowed them to analyze the general properties of human DTs, their construction, and their application as tools in the design of intelligent technologies [34]. Subsequent authors discuss the concept of DT and human-machine collaboration. They presented technologies and methods supporting the creation of human-oriented applications based on DT, including from the perspective of Industry 5.0. They analyzed the state of information resources related to DT in human-machine applications, with particular emphasis on the human perspective [35]. A digital twin of the human body is a virtual representation of a person's physiological state, created in real time from sensors and medical testing devices. The purpose of the representation is to simulate, predict, and optimize health outcomes through advanced analysis and modeling [36]. Human body DT has the potential to revolutionize healthcare. In this context, the authors of one study reviewed the status and prospects of human body DT and proposed a five-level development plan, starting from sensing components and ending with systems for acquisition, analysis, and decision-making [37]. The article also highlights the necessary support, safety, cost, and ethical issues that must be considered to ensure responsible and effective implementation of human body DT. They defined a framework for the development of human body DT, facilitating new interdisciplinary research and innovative solutions [36]. To summarize this brief review, the issues related to the construction and operation of DT can be divided into the following topics: purpose of application, conceptual elements, data links, model accuracy, interface, synchronization, and input data. The purpose of application is defined in the context of a real-world object [14]. In this approach, DT can be used to optimize processes within a real-world object or as a data processing and archiving center [14].

The connection between instances in DT can be one-to-one or based on independent mapping. In the latter case, virtual mapping can be a separate entity from the real system. Data links characterize communication between DT and its real counterpart. There are two possibilities: a one-way data link or a two-way link. The former allows data to flow from the physical object to the virtual one. The latter, in addition to the data received by the DT, also sends data to the object. In this approach, the DT interacts with the physical object, but can, for example, replace its parts [14]. Accuracy of representation means that the DT is an accurate representation of reality and describes every detail (at a given level). Partial accuracy means that the DT focuses only on the most important elements of the real object. The interface determines the ability to transfer data in the DT-PO system. A distinction is made here between human-machine and machine-machine interfaces [14]. Synchronization is another property of DT. Synchronization occurs when data is constantly exchanged between the real system and its image. Lack of synchronization occurs when such data exchange does not function. Input data is transmitted via data links. A distinction is made here between raw data and processed data [14].

2 Materials and Methods

This article presents only the structure of the DT model and the individual components and characteristics processed within its operation. Detailed research results using the DT will be presented in subsequent publications. This study proposed a simple DT model for an off-road vehicle for people with special needs (ATV). The fundamental problem posed in the context of such a DT is the study of vehicle-road interactions. Interactions resulting from the ATV's suspension, design, and drive type are examined. These vehicles are intended for use on mountain trails. Therefore, the terrain is uneven, causing these vehicle-road interactions to be dynamic, often with significant impact on the driver. As these vehicles are designed for ATVs, this is a significant issue. Furthermore, in the case of foot or manual drive, additional interactions are introduced in the vehicle's measurement system. However, this is not analyzed in our case due to the vehicle's design, which does not include (at this stage of research) a mechanical drive; we are using only an electric drive. The vehicle design is presented in two views in Fig. 1.

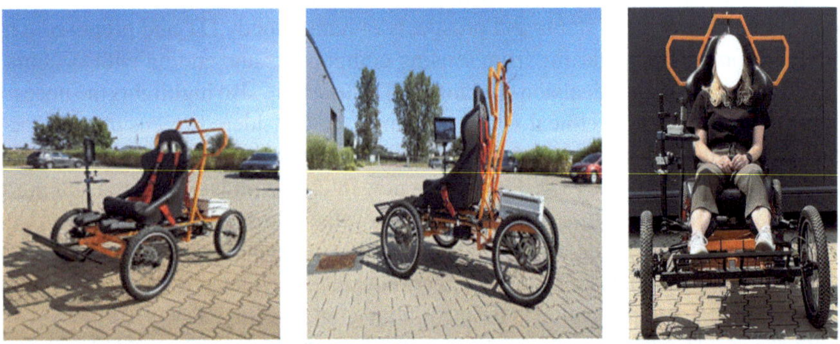

a b c

Fig. 1. Off-road vehicle for PWSN: a/ angle view b/ side view c/ front view. Source: photo by M. Staniek

To investigate the interaction in the vehicle-road system (mountain trail), several measurement elements were used. First, the current vehicle position was recorded using GPS and derivatives. This notation allows for assigning the recorded measurement data within the DT data structure to a specific cross-section on the mountain trail. It is important to remember that the variability of mountain trail cross-sections is significantly greater than that of a typical public road. Two different receivers were used in the study. The first is a GY-NEO6MV2 GPS module. This module has an external antenna. It is used to determine the position of the PWSN vehicle based on the GPS/Glonass signal. Communication with the controller takes place via a UART interface at a speed of 9600 baud. GPS records the PWSN vehicle's location, speed, altitude, and real-time clock

(RTC). The modules used are low-voltage and consume little power [37]. The GPS module provides auxiliary information such as ephemeris, almanac, estimated last position and time, and satellite status, as well as an optional time synchronization signal [37]. The second receiver is the Gravity DFRobot TEL0157. This is a GNSS/GPS/BeiDou receiver module with a Gravity connector manufactured by DFRobot. The GNSS positioning module supports both single and multiple satellite positioning systems. The GPS system for PWSN is presented on the Fig. 2.

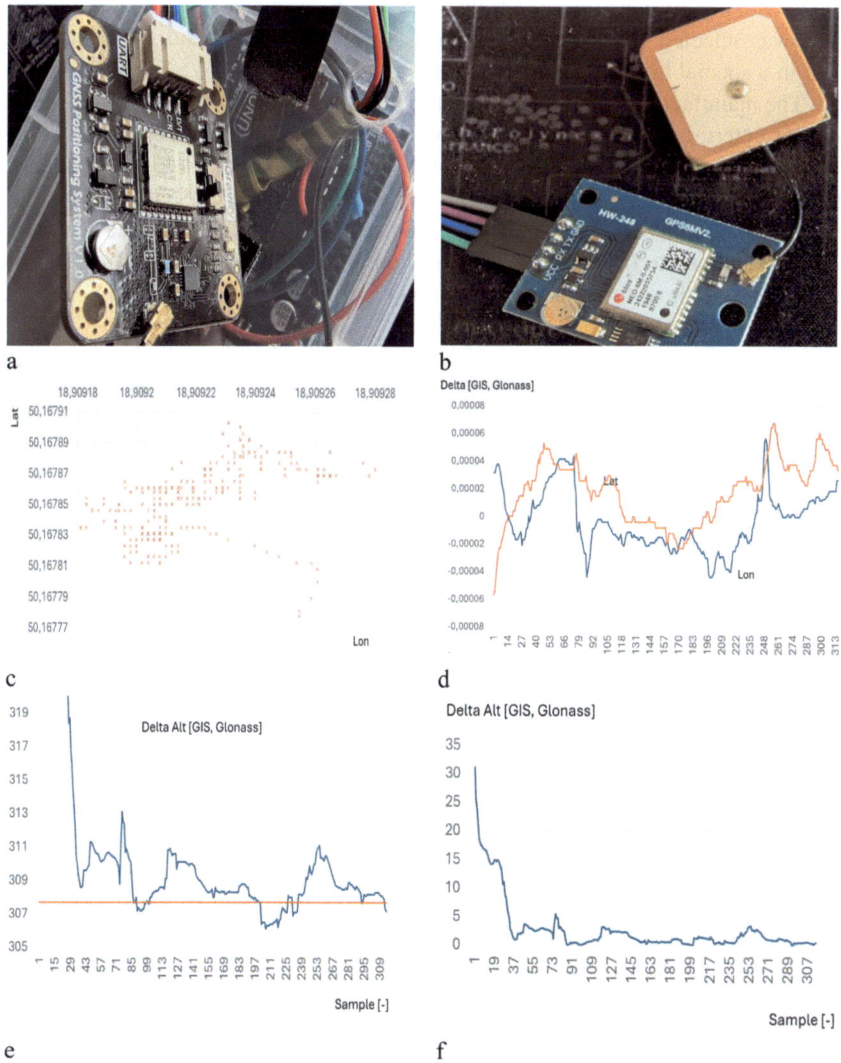

Fig. 2. GPS system for PWSN: a/ Glonass module, b/ GPS module, c/ tested accuracy data, d/ fixed position reading difference, e, f/ difference and error in reading the set height

For both receivers, location accuracy was tested based on location references and altitude above sea level according to GIS data. With average errors in latitude and longitude of 1.80E-05; 2.39E-05, the positioning accuracy was 1.901 m in longitude and 2.658 m in latitude. This means that the receivers are consistent with the manufacturer's specifications. The average difference in altitude readings was 2.793 m, which is a very good result. It can therefore be assumed that the receivers will position the PWSN vehicle on the mountain trail within its width.

To measure the PWSN vehicle's interaction with the ground, linear accelerometer modules with a gyroscope (ranging from 3 to 8) were installed. A 6-axis motion sensor module was selected. The module integrates the accelerometer and gyroscope and is housed in a special housing enabling flexible mounting to the PWSN trolley's support frame. The manufacturer's module designation is DFRobot SEN0386 [38]. The module allows for determining the object's position in space, measuring acceleration rate, and angular velocity. The measurement range for linear acceleration is ± 2.4, 8, and 16 g, respectively, and for angular velocity: ± 250, 500, 1000, and 2000°/s [38]. The digital filtering technology used in this module eliminates noise and increases the accuracy of the results obtained to 0.05 degrees. It is powered by a voltage ranging from 3.3 V to 5 V. The sensor's mounting locations are shown in Fig. 3.

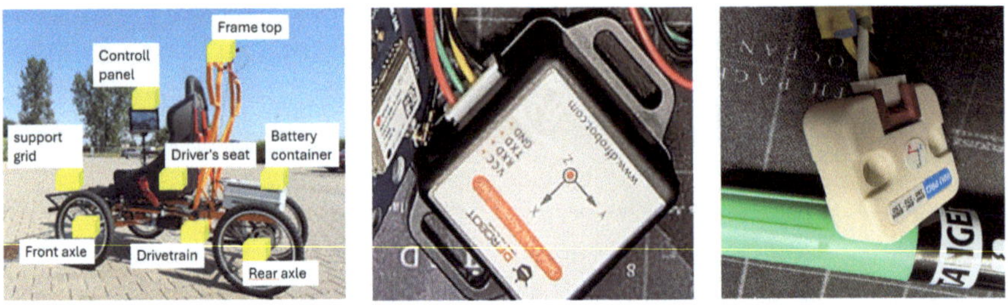

a b c

Fig. 3. Selected vehicle subsystems: a/ Assembly locations for SEN0368 on off-road vehicle for PWSN, b/ view of the sensor with mounting brackets, c/ atmospheric pressure, altitude field sensor

The BMM150 sensor (Fig. 3c) enables precise navigation by measuring magnetic fields. Integrated into the same housing, the BMP280 provides data acquisition for atmospheric pressure and altitude. This sensor is connected to a separate controller. In the prototype tests, the sensors are mounted on cable ties, as they have special mountings (Fig. 3b, c). Special cable ties allow for very stable mounting of the sensors, which is important for accelerometers. Mounting locations are shown in Fig. 3a. Acceleration measurements are taken for the following wheelchair components: both axles, the driver's seat, the drive, the leg cage, the upper frame, and the control panel. The sensor configuration for each test (from 3 to 8) is determined for each test. These mounted sensors provide information on the linear and angular acceleration experienced by the vehicle user and its components/elements. Both types of modules mentioned (GPS, ACC, BPM) are integrated with the controller(s) based on the Arduino UNO/Leonardo/MEGA2560 development board.

However, they can be integrated with any other system that supports this measurement system design standard, e.g., ESP32/8266, etc. In the longer term, the system will be fully implemented using a Raspberry Pi Zero 2W board. This system enables data exchange using Bluetooth and WiFi [39]. Currently, data is collected from the sensors directly to a PC via USB or an SD card. The use of Bluetooth/WiFi/LoRa will enable data transmission within the range of WiFi or another network, i.e., a maximum of 10/20/100 m to several kilometers. Ultimately, transmission of selected data describing the state of the measurement system will be implemented using the authors' proposed LoRa-based communication system, dedicated to the PWSN vehicle [40]. Here, depending on the needs, transmission in mountainous terrain can be achieved from one to several kilometers, depending on the location of the measurement base (this is a separate research problem). The authors tested the LoRa signal and obtained a continuous data flow at a guaranteed distance of up to 1.5 km using the 433 MHz standard [40]. The view of the sensor and sample data are presented in Fig. 4.

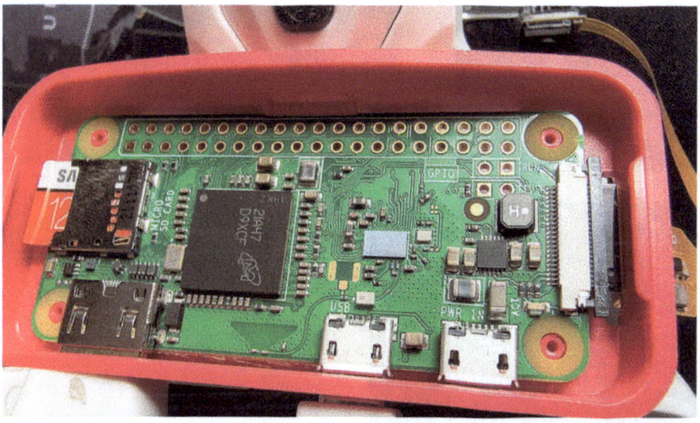

a).

b).

Fig. 4. Other measurement and telecommunications systems a). Raspberry Pi2W controller, b). LoRa prototype

The prototype shown in Fig. 4a, based on Raspberry Pi Zero 2W, is currently being implemented. It will enable communication with the traffic control center for PWSN. In addition to mapping the vehicle's location and the forces acting on it, it is equipped with devices that record data, including objects located in front of the PWSN vehicle. The RPLIDAR A2M12 [41] is used for this purpose. It is used as a distance sensor with 360° detection. The radar is supplied by Slamtec [41]. The RPLidar A2M12 scanner allows scanning from 8k to 16k samples per sec. in the range from 20 cm to 1200 cm. This enables sufficiently accurate examination of the position of various objects, including obstacles in front of the vehicle for PWSN. Mountain trails are usually narrow. Therefore, such a measurement allows for the comparison of data obtained in the accelerometer system with objects occurring on the mountain trail. It is possible to adjust the frequency of the scanner, and the angular resolution is 0.225° [41]. The view of obstacle detection system is presented on the Fig. 5.

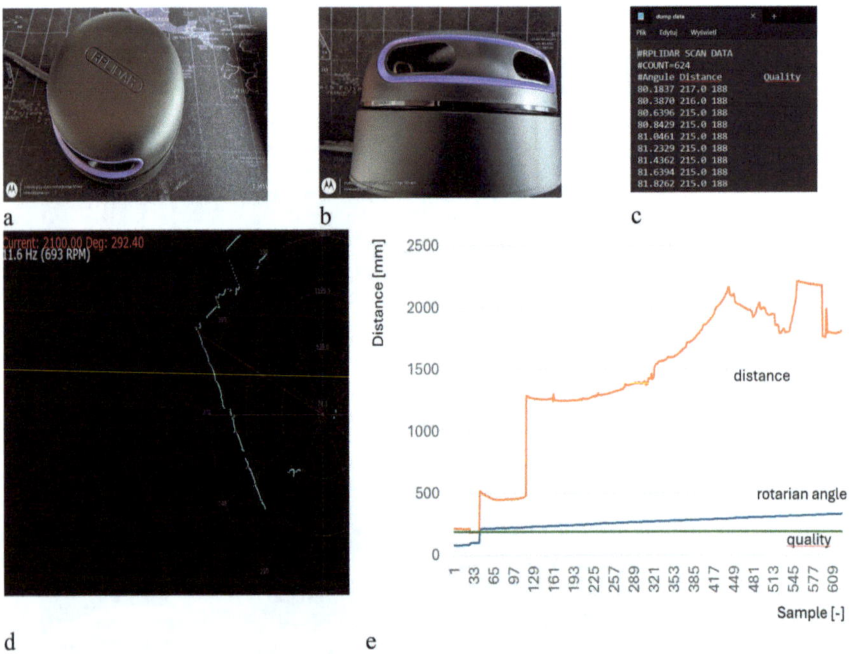

Fig. 5. Obstacle detection system: a, b/ Scanner A2M12, c/ dump data, d/ frame grabber for A2M12, e/ data graph

In addition, to detect objects in the scene of movement on a mountain trail and associate them with other data structures in the DT system, the PWSN person may be equipped with a proprietary optical tracker. This is a device that integrates two USB cameras on a special head-mounted mount. One camera is responsible for recording the movement of the left, right, or both eyes of the driver. The other camera is responsible for recording objects in the scene. The measuring device is shown in Fig. 6.

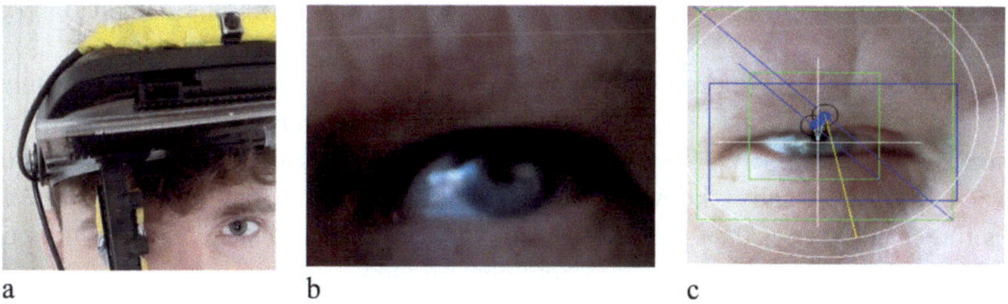

Fig. 6. Eye tracker for PWSN: a/ tracker device, b/ view from eye camera, c/ eye movement analysis in Authors application

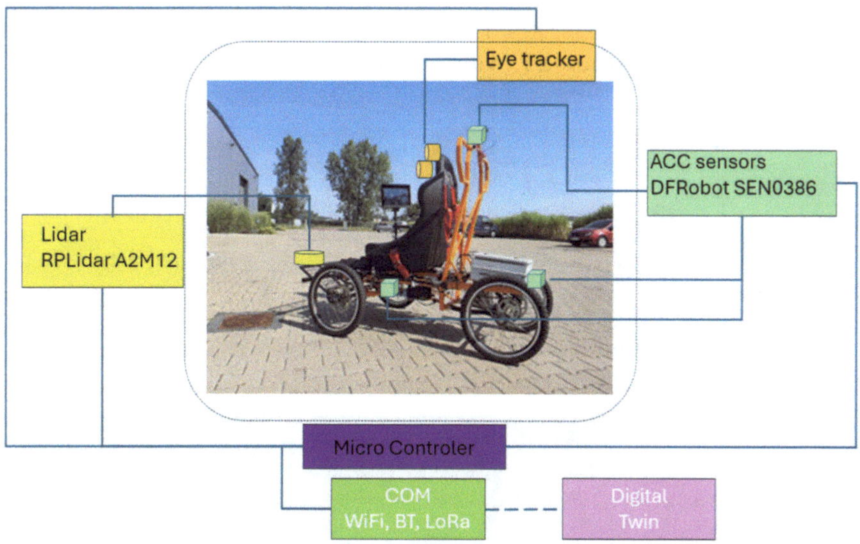

Fig. 7. Measurement system for off-road vehicles for PWSN

The device shown in Fig. 6 uses two USB cameras, which do not require additional drivers in a Windows environment. This module has now been replaced with a newer, very lightweight and user-friendly one. The first OV5693 5MPx USB camera is compatible with Raspberry Pi and Jetson Nano. The manufacturer's code is Waveshare 24710 [42]. The camera offers a 5MP resolution and a 135° field of view. The camera is also equipped with autofocus, and the supported recording formats are MJPG and YUY2. The image recording rate is 15 FPS and 2592 × 1944 [42]. The second camera, the ArduCAM5MP, is also an autofocus camera with a 5MP resolution. The maximum resolution and frame rate are 2592 × 1944 at 30 fps. Both cameras are compatible with various platforms and operating systems, such as Windows, Linux, Mac, and Android. The manufacturer's code for the second camera module is ArduCAM B0441 [43]. The result is a system that allows for the analysis of data related to the movement of an off-road vehicle for the PWSN on a mountain trail. The entire measurement system is presented in the diagram in Fig. 7.

Figure 7 shows the entire DT system. It consists of four measurement subsystems: vision, vibroacoustic, lidar, and GPS. The digital twin of our design, to use colloquialisms, sees what the PWST driver sees (and often more), senses much more than the driver (more analytically), knows where the vehicle is located, and identifies the vehicle's foreground better than the driver (although at a shorter distance at the moment). The Fig. 8 shows the measurement system general view.

The following figures present sample data from the measurement subsystems. Figure 9 illustrates data from the GPS subsystem. Figure 10 illustrates data from the accelerometer subsystem. Figure 11 shows data from the Lidar subsystem. Figure 12 shows data obtained from the eye tracker subsystem.

As shown in Figs. 9, 10, 11 and 12, our DT enables the collection of real-time data on the condition and movement of the vehicle for PWSN across a very wide range. Our twin sees, hears, detects, and locates. This allows for full control using communication in one of the popular standards: BT, WiFi, or LoRa.

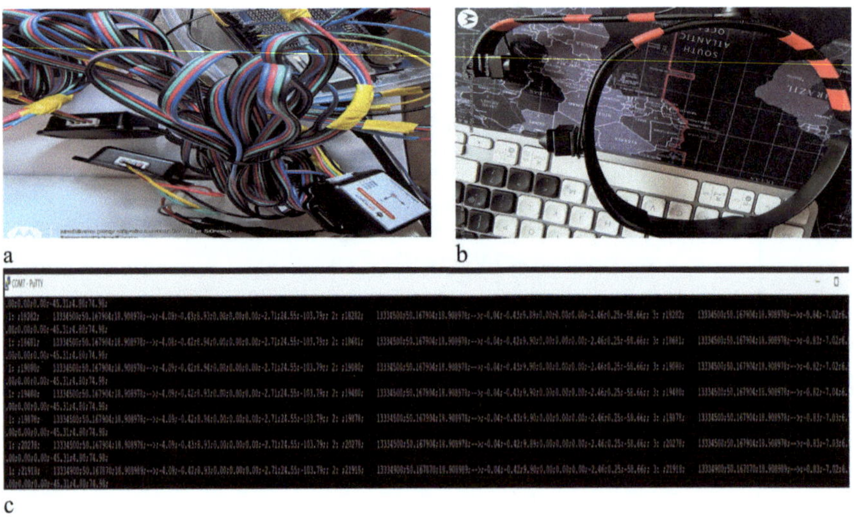

Fig. 8. Measurement system general view: a/ controller side, b/ new version of ET c/ sensor data in Putty

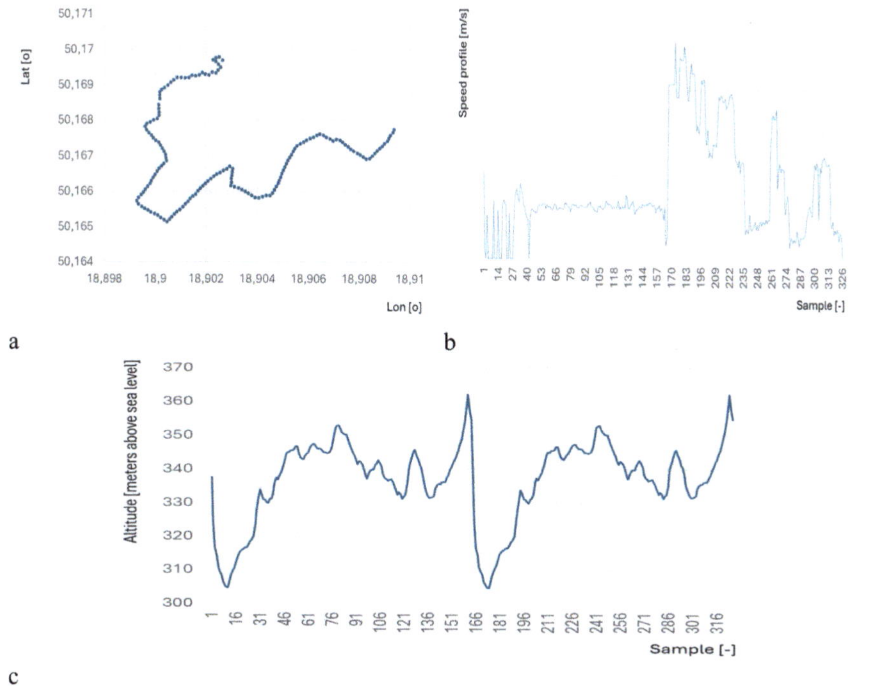

Fig. 9. Example data in GPS subsystem: a/ vehicle movement path, b/ recorded vehicle speed PWSN, c/ recorded altitude above sea level

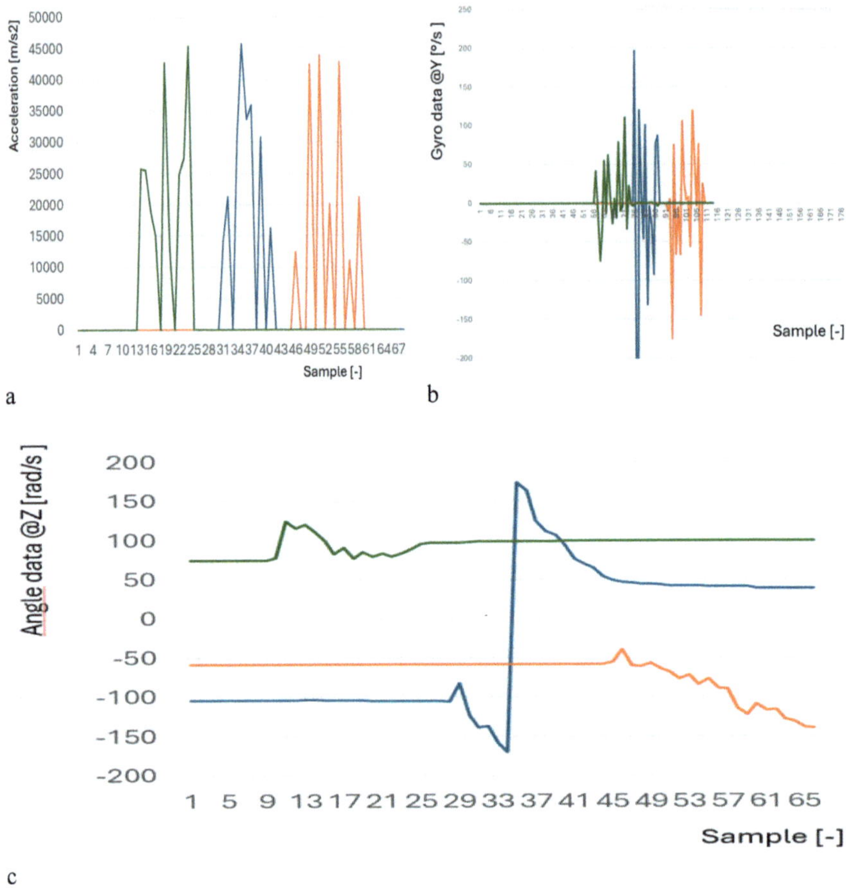

Fig. 10. Example data in accelerometer subsystem: a/ acceleration from 3 sensors in x axis, b/ data from Gyroscope @ Y axle, c/ data from angle at Z axle

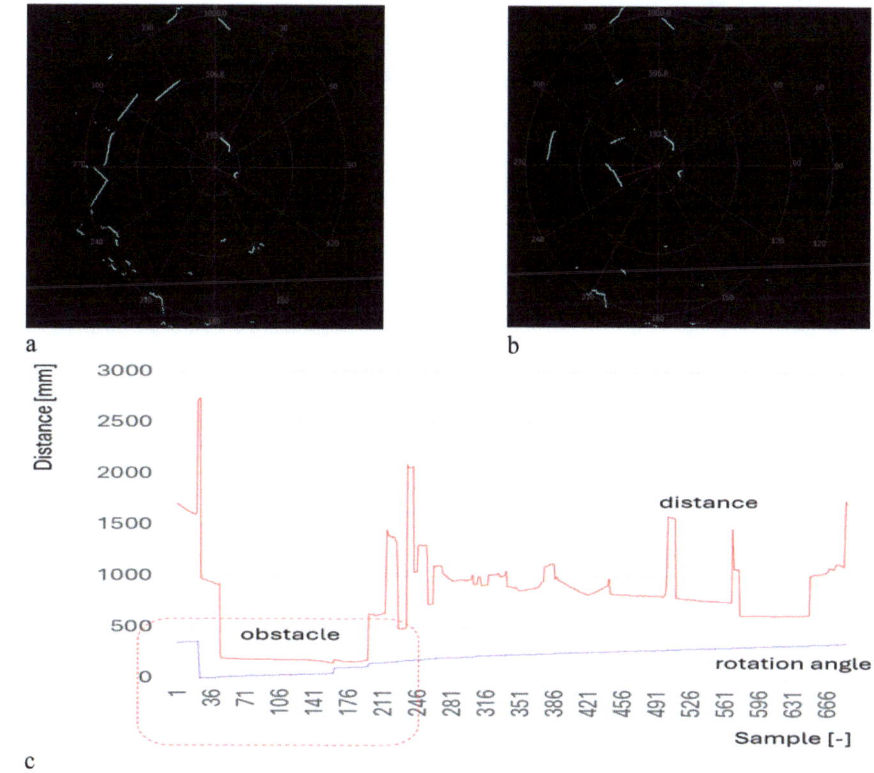

Fig. 11. Example data in Lidar subsystem: a/ open space, far distance obstacle, b/ near distance obstacle, c/ dumpdata

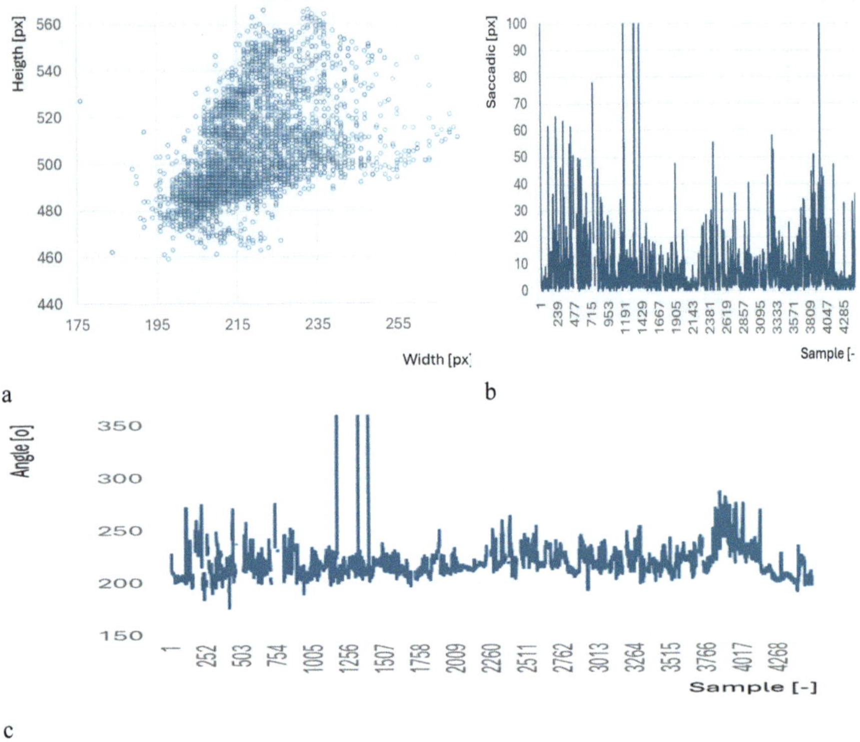

Fig. 12. Example data in eye tracker subsystem: a/ fixation point, b/ saccadic movement, c/ angle rotation

3 Component

The data flow within the DT framework between the real object and the model is presented below. This data flow allows for the implementation of an interaction model in the PWSN vehicle-mountain trail system. This is a basic model that examines the behavior of the prototype vehicle on a mountain trail. It allows for modifications to the PWSN vehicle's structural elements. Due to editing constraints, this model will be described in detail in another publication. The interaction model (IM) can be described as an ordered seven-step model:

$$IM = < L,\ A,\ G,\ D,\ O,\ ET,\ MTR > \quad (1)$$

where: L – location data set (GPS), A- linear acceleration data set (ACC), G - angular acceleration data set (GYRO), D - set describing rotation angles relative to the axis (Angle), O- LIDAR data set, eye tracker data set, MTR- GIS reference set for cross-sections on the mountain trail.

The Fig. 13 shown the data flow in DT.

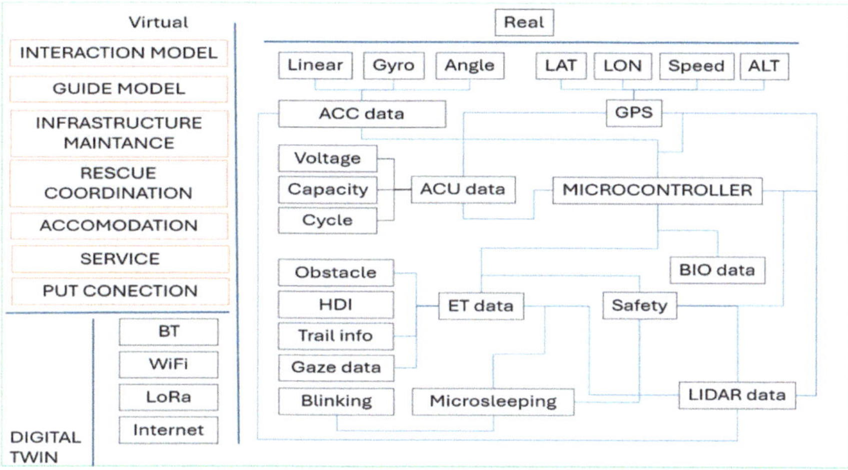

Fig. 13. Data flow in DT

Other models are also planned to be implemented as part of DT. The guide model (GMODE) is responsible for interactively guiding the PWSN vehicle on the mountain trail. The infrastructure maintenance model (IMAOD) is responsible for scheduling modernization, repair, and construction work on the mountain trail based on data obtained from PWSN rides. The rescue coordination model (RCMOD) is based on data collected from vehicles. The accommodation model (ACCOM) is based on the current locations of PWSN vehicles on the route and the location of the nearest tourist infrastructure points, including infrastructure for PWSN, and provides accommodation for PWSN passengers. A service module (SERVI) for virtual vehicle diagnostics and minor repairs on the route and major repairs off the route. The model is based on the location of repair stations and charging points for PWSN wheelchairs on the route. A module for connecting PWSN with public transport (PuTOD), allowing real-time coordination of PWSN traffic with train and bus timetables. Other modules not described in detail here concern:

- obstacle, a module for detecting obstacles and hazards on the trail (ET and LIDAR),
- HDI, a module for analyzing the open telemetry interface of the PWSN wheelchair,
- safety, a module analyzing safety on the trail, e.g., related to falling asleep while driving,
- BIO module, collecting basic biometric data from the user, including heart rate, temperature, EMG activity, ECG, etc.

This biometric module will be designed using low-cost biometric sensor and microcontroller components. The cost of the entire DT is being designed using low-cost architecture, ensuring its production cost is a fraction of the price of a PWSN wheelchair. The cost of all DT components in the full version is approximately €655 (8 accelerometers, LIDAR, GPS, and others). After sensor integration, the cost should drop to €400, and without LIDAR, to €150.

A discussion of the results obtained using individual measurement modules will be included in subsequent publications. This publication is intended to present the DT concept for PWSN and the technologies used in its implementation.

Ultimately, the DT will consist of a model that not only integrates the PWSN vehicle with the mountain trail but also with the PWSN driver. For this purpose, a PWSN driver model (DTHM_PWSN) will be formulated, considering the specificity of the driver's diagnosed conditions (conditions for mountain wheelchair operation). For this purpose, biometric devices such as a heart rate monitor, EEG, ECG, and EMG will be mounted on the PWSN driver's body. The Plus will be configured appropriately for this ET examination. Due to the cost of the EEG cap, it will be the only DT instance for this type of analysis. As indicated by the description above, the PWSN twin project is still in development and will be presented in a series of publications.

4 Conclusions

The DT model allows for realistic representation of data acquired by a person with PWSN through various senses while traveling along a mountain trail. The ET subsystem allows for the use of computer vision procedures. Here, in a broader sense, the image is processed contextually, differently than a human sees it and often for other purposes. The vibroacoustic subsystem collects information about the forces experienced by a person with PWSN traveling in a wheelchair, as well as the forces acting on the structural elements of the wheelchair. This makes it possible to examine the quality of the seat (riding comfort), the quality of the leg rest attachment, the quality of data acquisition using the PWSN terminal, the method of vibration damping in the drive and suspension systems, but above all to examine the interaction of the vehicle with the trail. The latter can be used to determine the scope and schedule of repair, construction, and modernization works on the mountain trail, e.g., widening the trail, leveling unevenness, slopes, etc. It should be remembered that each PWSN vehicle equipped with our measurement system is a DT instance on the trail, and each of them collects new data in similar but also different contexts.

All data passing between the physical and virtual objects is located at specific cross-sections of the mountain trail using GPS positioning. Communication in the opposite direction, between the model and the physical object, allows for near-real-time guidance for the PWSN driver. For example, it can assist with navigation on a difficult mountain trail, avoid obstacles, suggest instantaneous operating parameters for the drive, and so on.

Acknowledgements. The present research has been financed through the *National Centre for Research and Development* as a part of the competition within the scope of "Things are for people" in project with title: "Integrated platform for planning, organization, supervision and support for the availability of mountain tourism offer for people with difficulties in physical functioning and a specialized off-road vehicle for the implementation of the tourism offer - Mountains without barriers" realized by Silesian University of Technology, Gliwice.

Rzeczpospolita Polska
Dostępność Plus
NCBR
Politechnika Śląska

References

1. Matchett, R., Wium, J.: Digital twins for road infrastructure. https://repository.up.ac.za/bitstream/handle/2263/87359/
2. Wynand, S., Broekman, A.: Development of a digital twin of a local road network: a case study. J. Testing Eval. **51**, 1–15 (2021)
3. Crampen, D., Blankenbach, J.: LOADt: towards a concept of level of as-is detail for digital twins of roads. https://www.ucl.ac.uk/bartlett/construction/sites/bartlett_construction/files/1723.pdf
4. Oussama, M., Taleb, T., Seung. S.J.: Roads infrastructure digital twin: a step toward smarter cities realization. IEEE Network, 1–8 (2020)
5. Rudskoya, A., Ilina, I., Prokhorovb, A.: Digital twins in the intelligent transport systems. Transp. Res. Procedia **54**, 927–935 (2021)
6. Jiang, F., Ma, L., Broyd, T., Chen, W., Luo, H.: Building the digital twin of existing highway using map data based on the engineering expertise. https://discovery.ucl.ac.uk/id/eprint/10140121/1/AUTCON-D-21-00424_R1%20copy.pdf
7. Vieira, J., Poças Martins, J., Marques de Almeida, N., Patrício, H., Gomes Morgado, J.: Towards resilient and sustainable rail and road networks: a systematic literature review on digital twins. Sustainability **14** (2022)
8. Shamlitsky, Y., Ovsyankin, A., Morozov, E., Strekaleva, T.: Using digital twins to manage traffic flows. In: E3S Web of Conferences, vol. 471 (2024)
9. Consilvioa, A., et al.: Towards a digital twin-based intelligent decision support for road maintenance. Transport. Res. Proc. **00**, 1–8 (2022)
10. Thonhofer, E., et al.: Infrastructure-based digital twins for cooperative, connected, automated driving and smart road services. https://elib.dlr.de/195083/1/OJ-ITS-2022-12-0079-21-12-2022.pdf
11. Schaniel, J.: Towards a Digital Twin of Seychelles. Road Transport System https://ethz.ch/content/dam/ethz/specialinterest/usys/tdlab/docs/education/theses/2022-msc-schaniel.pdf
12. Callcut, M., Cerceau Agliozzo, J.P., Varga, L., McMillan, L.: Digital twins in civil infrastructure systems. Sustainability **13** (2021)
13. Sanfilippo, F., Thorstensen, R.T., Jha, A., Jiang, Z., Robbersmyr, K.G.: A perspective review on digital twins for roads, bridges, and civil infrastructures. In: International Conference on Electrical, Computer, Communications and Mechatronics Engineering, Maldives, pp. 1–6 (2022)
14. Kochan, A.: Cyfrowy bliźniak systemu sterowania ruchem kolejowym. Zeszyty Naukowo-Techniczne SITK RP, Oddział W Krakowie, Nr 2(121), 2020 (in polish)
15. Piromalis, D., Kantaros, A.: Digital twins in the automotive industry: the road toward physical-digital convergence. Appl. Syst. Innov. **5** (2022)
16. Krishnamurthy, P.: Digital twin system for automotive safety with integration of a safety controller (2024)
17. Yalavarthy, U.R.S, Kumar, B., Babu, A,R,V., Narasipuram, R.P., Padmanaban, S.: Digital twin technology in electric and self-navigating vehicles: readiness, convergence, and future directions. Energy Convers. Manage. **26** (2025)

18. Gao, J., Peng, C., Yoshinaga, T., Han, G., Guleng, S., Wu, C.: Digital twin-enabled internet of vehicles applications. Electronics **13**(7) (2024)
19. Polat, A.O., Erden, B.C., Kul, S., et al.: Light electric vehicle performance with digital twin technology: a comparison of motor types. Arab. J. Sci. Eng. **49**, 7209–7222 (2024)
20. Popa, L., Berdich, A., Groza, B.: CarTwin - development of a digital twin for a real-world in-vehicle CAN network. Appl. Sci. **1** (2022)
21. Kabir, R., Ravi, B.B.Y.: RAYS.: Digital twin technologies for vehicular prototyping: a survey. Intell. Transp. Syst. 6 (2025)
22. Pfeiffer, J., et al.: Modeling Languages for Automotive Digital Twins MODELS '24, September 22–27, Linz, Austria (2024)
23. Heithoff, M., Konersmann, M., Michael, J., Rumpe, B., Steinfurth, F.: Challenges of integrating model-based digital twins for vehicle diagnosis. In: International Conference on Model Driven Engineering Languages and Systems Companion, pp. 470–478, IEEE, Västerås, Sweden (2023)
24. Spiryagin, M., Edelmann, J., Klinger, F., Cole, C.: Vehicle system dynamics in digital twin studies in rail and road domains. Veh. Syst. Dyn. **61**(7), 1737–1786 (2023)
25. Tran, D., et al.: Advanced digital twin framework for electric truck. In: IEEE Vehicle Power and Propulsion Conference (2023)
26. Ju, S., van Vliet, P., Arenz, O., Peters, J.: Digital twin of a driver-in-the-loop race car simulation with contextual reinforcement learning. IEEE Rob. Autom. Lett. Preprint Version (2023)
27. Genc, M.O.: Development of a digital twin model for electric bikes based on urban experimental driving in full electric mode. J. Mech. Sci. Technol. **37**, 4349–4359 (2023)
28. Haralampous, O., Samaras, N.: Digital twin of an electric bicycle using smartphone sensors and weather data. In: Nathanail, E.G., Gavanas, N., Adamos, E. (eds) Climate Crisis and Resilient Transportation Systems. CSUM 2024. Lecture Notes in Intelligent Transportation and Infrastructure. Springer, Cham (2024)
29. Rana, A., Petitti, A., Ugenti, A. Galati, R., Reina, G., Milella, A.: Toward digital twin of off-road vehicles using robot simulation frameworks. IEEE Access, 1 (2024)
30. Robyns, S., Heerwegh, W., Weckx, S.: A digital twin of an off highway vehicle based on a low cost camera. Procedia Comput. Sci. **232**, 2366–2375 (2024)
31. Cheng, Z.: Human digital twin with applications. In: Proceedings of the 7th International Digital Human Modeling Symposium, vol. 7 (2022)
32. Wolfgang, M., Lauer-Schmaltz, Cash, P., Hansen, J.P., Maier, A.: Towards the human digital twin: definition and design - a survey. arXiv:2402.07922 (2024)
33. Soman, R. K., Farghaly, K., Mills, G., Whyte, J.: Digital twin construction with a focus on human twin interfaces. Autom. Constr., 170 (2025)
34. Saariluoma, P., Myllylä, M., Karvonen, A., et al.: A human digital twin for the M-Machine. Discov. Artif. Intell. **4**, 61 (2024)
35. Krupas, M., Kajati, E., Liu, C., Zolotova, I.: Towards a human-centric digital twin for human–machine collaboration: a review on enabling technologies and methods. Sensors **24**(7) (2024)
36. Tang, Ch., Yi, W., Occhipinti. E., Dai, Y., Gao, S., Occhipinti, L.G.: A roadmap for the development of human body digital twins (2024)
37. U-blox module. https://content.u-blox.com/sites/default/files/products/documents/NEO-6_DataSheet_%28GPS.G6-HW-09005%29.pdf
38. DFRobot SEN0386. https://wiki.dfrobot.com/Serial_6_Axis_Accelerometer_SKU_SEN0386
39. Raspberry Pi Zero 2W. https://www.raspberrypi.com/products/raspberry-pi-zero-2-w/
40. Staniek, M., Celiński, I.: The use of LoRa technology as an alternative to GPS in the navigation of a mountain vehicle intended for people with special needs. Archives of Transport, Warsaw (2025) (under review)

41. RPLidar A12m12. https://www.dfrobot.com/product-1461.html
42. Waveshare OV5693. https://www.waveshare.com/ov5693-5mp-usb-camera-a.html
43. Arducam. https://www.arducam.com/5mp-autofocus-usb-camera-module-with-single-microphone-for-windows-linux-macos-android

Economic, Regulatory, and Social Aspects

Involvement of Pension Funds in Public Private Partnership Projects

Václav Baroch(✉) and Rudolf Franz Heidu

Faculty of Transportation Sciences, Czech Technical University in Prague, Konviktská 20, Prague 1 110 00, Czech Republic
{barocvac,rudolf.heidu}@fd.cvut.cz

Abstract. Currently, the EU is facing an ageing population. The key task of EU Member States is therefore to ensure a sustainable pension system. One of the tools used for this sustainability is long-term savings in pension funds. In order for this tool to be functional, the pension fund must ensure the preservation of the value of savers' savings over the long-term savings horizon. Deposits in pension funds must be stored safely and be resistant to negative fluctuations in financial markets. The article deals with the idea of interlinking Pension Funds with Public Private Partnership projects in the field of transport. This connection has the potential to bring Pension Funds the sought-after safe investment, which also has to overcome inflation in the long term. The use of accumulated capital of Pension Funds in public PPP projects, which can simultaneously bring a rejuvenating financial injection and increase the efficiency of competitive projects.

Keywords: Public Private Partnership projects · Pension Funds · pensions · profitability

1 Introduction

Funding models in the pension system are a highly debated topic. The discourse is mainly focused on social old-age security, i.e. the so-called Pillar I of old-age security in the Czech Republic. Pillar II is defined as supplementary pension insurance with capital accumulation and mainly private fund management.

At the same time, however, the pension fund must ensure a return that at least matches inflation. Conservative investments in money markets are stable and safe, but they usually do not cover inflation. In contrast, investment strategies in stock markets outperform inflation in the long term, but at the same time they are associated with increased risk in the event of negative market developments.

An important question is whether the ongoing financing is prospective and sufficient despite the negative demographic development. In addition, the so-called Mackenroth thesis on the macroeconomic equivalence of funding models for pension expenditure (i.e. funded and pay-as-you-go) over time has been formulated in German scientific discourse.

E. Macioszek et al. (Eds.): TSTP 2025, LNNS 1769, pp. 197–209, 2026.
https://doi.org/10.1007/978-3-032-14094-4_13

New forms of financing for public transport works could expand both basic forms. Private-public cooperation and the linking of old-age provision and investment in infrastructure construction could generate a pension system 4.0 with considerable crisis resistance.

Possible alternative forms of PPP (Public-Private Partnership projects) combined with pension capital open up the possibility of a robust alternative investment strategy for pension funds.

The article examines the current financial, legislative and social basis for the involvement of Pension Funds in PPP projects. New forms of financing public transport projects could expand both basic forms. Cooperation between the private and public sectors and the connection of old-age provision.

Part 2 analyses the current situation in the Czech Republic. Part 3 then lists possible alternatives for financing Pension Funds. Part 4 presents PPP projects as a significant alternative and presents practical experiences from abroad. Part 5 develops a discussion on the benefits of PPP projects for Pension Funds. Part 6 then analyses the financial potential of Pension Funds' capital for investments in PPP projects in the Czech Republic.

2 Brief Analysis of the Market in the Czech Republic

2.1 Underperformance of Pension Funds

There are 2 types of funds on the market in the Czech Republic today, Participating and Transformed, with only the Participating type of fund available to new clients.

The Transformed type of fund is completely conservative, and it is considered that the participant practically cannot lose the saved amount, he/she can only gain returns (if the pension savings system as such does not collapse and the Pension Fund company itself persists in its activity).

This security is redeemed by a very low interest rate that does not beat inflation in the long term.

Economists have been pointing out for years that transformed funds devalue participants' savings under current conditions and have recommended switching to participant funds that offer higher returns as a remedy for this devaluation.

To a lesser extent, these recommendations emphasise that higher returns are associated with higher risk and, above all, that participants may lose a substantial part of their savings in the event of adverse market developments.

Participating funds offer a fundamental change from the transformed fund in the choice of investment strategy. While the choice of a conservative/hedged investment strategy in the Participation Funds is more or less similar to the Transformed Fund with a slight swing towards higher returns and slightly higher risk, the other strategies increase the risk of losing the investment in proportion to the higher returns.

In the current ever-present trend of strongly rising financial markets over the long term, investment advisers are overreacting by arguing that anyone who saves for retirement in a transformed fund, or a participating fund with a conservative strategy, is losing money and would do much better to switch to a fund with a bolder investment strategy.

Let us now leave aside the speculation that, when the trend of rising markets reverses and there is a deeper fall, articles will again be published assuring us that any right-minded person must have seen the fall coming.

Neither transformed funds nor participatory funds with a conservative strategy achieve returns that cover inflation in the long term and actually devalue clients' money in real terms.

Over the last 25 years (January 2000 to December 2024 to be precise), the long-term average rate of inflation expressed as the increase in the average annual consumer price index, which reflects the change in the average price level over the last 12 months compared to the average of the previous 12 months, is 3.19% (based on the analysis of data in [1]).

This is thus the inflation rate that pension funds should overcome if we want to have a theoretical inflation-type virtual EVA (Economic value added) at 0.

2.2 Model Example of Low Performance Pension Fund of Česká spořitelna, a.s.

Using the example of the returns of the transformed Pension Fund of Česká spořitelna, a.s., one of the largest funds in the Czech Republic, it can be demonstrated that the virtual EVA ratio relative to inflation is negative in the long term.

The long-term average return of the transformed Pension Fund of Česká spořitelna, a.s. over the last 25 years (2000–2024) is 2.10% (based on the analysis of data in [2]) (Figs. 1 and 2).

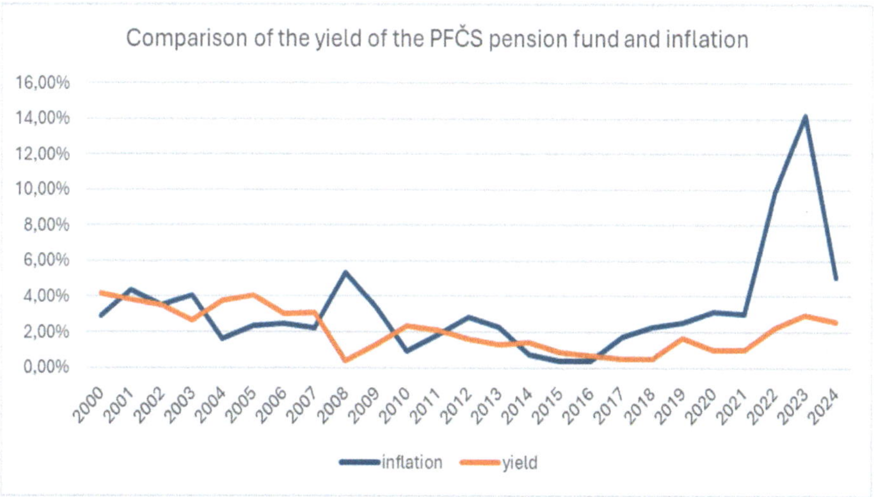

Fig. 1. Annual returns of the Pension Fund of Česká spořitelna, a.s. (Source: based on the analysis of data in [1, 2])

If we were to create a virtual EVA indicator based on inflation for these figures, it would take the following form with an overall average value of −1.24% per year.

Let us now disregard the externally supplied additional benefits in the form of the state contribution, tax relief and the nature of the pay-as-you-go investment, which has a variable return depending on the age of the participant, as these benefits are essentially independent of the profitability of the pension fund.

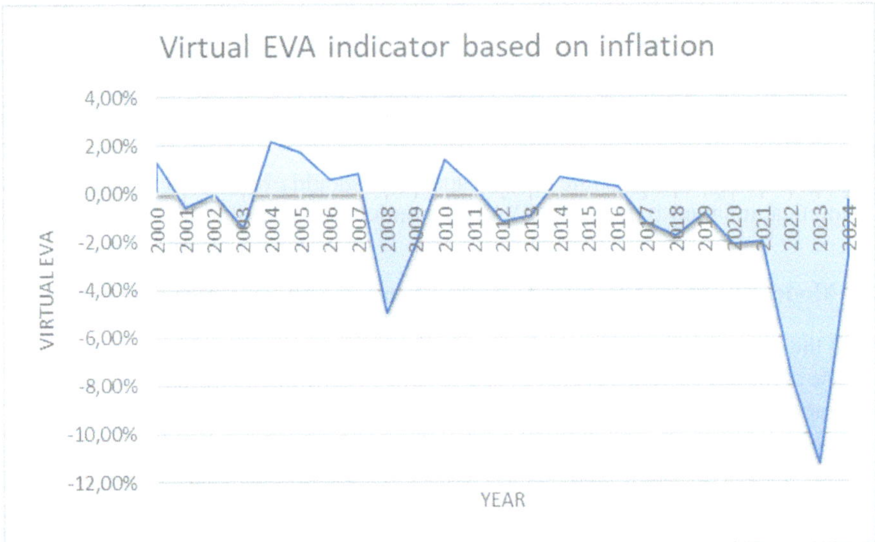

Fig. 2. Virtual EVA inflation/return of Česká spořitelna, a.s. Pension Fund. (Source: based on the analysis of data in [1, 2])

With an average virtual EVA of −1.24% per year, a depositor who had CZK 100,000 saved in the fund in 2000 now has a current value of only CZK 73,203 after 25 years.

The return of the transformed funds is therefore significantly loss-making in real terms. Participatory funds with a conservative strategy have not been around long enough to compare 25 years, but their returns in recent years show that they also devalue the funds invested by their participants in real terms.

2.3 A Broader Comparison of Pension Fund Performance

For the sake of comparison, let's compare internationally, with 2017 as a proxy, a period long after the Great Financial Crisis but well before the COVID period and the subsequent inflationary wave.

In this period, the real return of pension funds of −1,8% in the Czech Republic comes out as the worst of the comparative sample, which unfortunately confirms the above considerations. The average real return on pension funds was 3,67% in 2017 in selected OECD (Organisation for Economic Co-operation and Development) countries [3].

3 Alternative Safe Investment Opportunities

3.1 Existing Alternatives

An alternative for some of the dissatisfied savers may be participation funds with a bolder strategy, where especially the so-called dynamic strategies with a high degree of investment in shares have earned depositors above-standard double-digit returns in recent years, well above inflation.

However, bold investment strategies are not always appropriate for every participant. In the event of a sharp downturn in the financial markets, such strategies can quickly deprive participants of the returns they have earned. In the long term, the conservative investor looks in vain for a safe investment strategy that has the potential to beat inflation without the risk of sudden rapid loss of value. According to the data of the Ministry of Finance of the Czech Republic [4], at the end of 2024, there were a total of four million savers in supplementary pension schemes, evenly distributed in transformed funds (two million savers) and in Participation Funds (two million savers).

If we assume that ¼ of the Participation Funds' savers have chosen a conservative strategy, we arrive at a real number of 2.5 million savers whose deposits are devalued in real terms by the pension funds in the long term, which is partly subsidised by the state in the form of external benefits (mainly state contribution, tax rebate).

That being said, switching to the bold investment strategies of Participating Pension Funds are not suitable for everyone in all circumstances. In this sense, alternative options in the form of endowment life insurance or a general long-term investment product are not a solution due to either low returns or increased risk.

3.2 Variant of Combining Pension Funds and the Real Estate Market

The government's plan (Government of the Czech Republic February 2025) to allow pension funds to invest in housing, specifically investment construction and, alternatively, renting, seems very interesting.

The amount of the real impact of the benefit and the specific conditions are not yet known, but the plan may be limited by the generally problematic construction conditions in the Czech Republic, particularly the protracted permitting processes. This raises the concern that pension funds will not be able to invest more than units or at most lower tens of billions of CZK in housing in real terms.

4 PPP Projects in Transport Infrastructure

4.1 Participation of Pension Funds in PPP Projects in Transport Infrastructure

Another interesting option, which is the subject of this article, is the possibility of involving pension funds in PPP projects. PPP projects are collaborations between the public and private sectors to deliver public services or infrastructure projects. In these projects, the private sector undertakes financing, construction, operation or maintenance, while the public sector provides support, for example in the form of long-term contracts or contributions.

The main advantages of PPP projects include more efficient use of resources, risk sharing between sectors and the possibility of faster project delivery. On the other hand, these projects can be controversial due to higher costs or contractual complexity.

4.2 The Czech Railway Infrastructure Faces Many Challenges

Outdated technology: many lines still use older signalling systems, limiting capacity and safety. The transition to the modern European ETCS (European Train Control System) is in process but is costly and time-consuming to implement.

Signalling equipment failures: recent problems with ETCS failures have shown the vulnerability of the infrastructure. These outages led to significant train delays and service restrictions on major corridors. Insufficient investment: even when large investments are planned to upgrade lines, current funding is often insufficient to cover all needs. For example, an additional 4,000 km of lines are planned to be equipped with ETCS by 2030, requiring an investment of €32 billion.

Capacity problems: main corridors are often congested, causing delays and limiting the development of rail transport. The Czech Republic does not sufficiently exploit the potential of private capital in building public infrastructure. In Western Europe, nearly 2 000 private investment projects in public infrastructure worth €410 billion have been implemented over the last 30 years. For example, a total of 60 PPP projects have been implemented in railways with over €66 billion invested. The average contract duration for rail infrastructure projects is 35 years (lines + urban rail).

In recent years, the popularity of rail for long-distance and suburban transport has been growing again. Passenger traffic in the Czech Republic has increased by 70% over the last 10 years and with the advent of high-speed lines, further passenger shift to rail and the need for further high investment can be expected. Passenger transport is growing faster than expected [5]. The necessary investments will require new resources.

PPP projects have not been implemented much in the Czech Republic and there is therefore considerable potential. Expert studies indicate that the necessary legislation and environment for PPP projects already exist in the Czech Republic.

The likely reason for the low level of use of PPP projects to date is the need for high investment costs, where PPP projects are not suitable for ordinary small and medium-sized enterprises. Large companies usually have their own core business and are not very motivated to invest in public infrastructure in the Czech Republic.

4.3 Benefits of PPP Projects for the Investor

Outdated technology: many lines still use older signalling systems, The DBFM (Design-Build-Finance-Maintain) model appears to be appropriate for the involvement of pension fund capital in PPP projects. This is a model of public-private collaboration where the private partner designs, builds, finances and then maintains the infrastructure over a period of time.

This model is often used in the transport infrastructure sector, for example in the construction of roads, railways or bridges. This model allows an efficient allocation of risk between the public and private sectors and ensures the long-term quality of the infrastructure. Another model used is the DBFMO (Design–Build–Finance–Maintain-Operate) model, which adds the area of operation over the planned contractual life of the project.

Information on the real profitability of projects for the investor is mostly non-public. However, where we have been able to find at least some information, it seems that PPP projects represent a win-win instrument for both the contracting authority (usually the state, the region) and the investor (private capital).

The table shows that where some financial data on returns was available, there was a contractual agreement to index the payments to inflation or there were instruments to allow inflation to be considered in favour of the investor.

In addition to the direct financial revenues or computable returns there is an indirect, longer-term positive effect aimed directly at developers. Real estate development is possible especially at stations and in logistics areas and were railways border with great development potential. The scope of public service obligations (in passenger transport) needs to be addressed everywhere, as well as connections with road and/or urban transport networks [7]. This indirect effect is more general and cannot be applied locally to the specific projects concerned. However, the effect is present and its appropriate commercial and legislative capture further increases the potential for motivating market makers to develop PPP projects.

4.4 Risks Associated with the Involvement of Pension Funds in PPP Projects

Of course, PPP projects also involve some risk. One of the basic risks is the age structure of the promoters. According to data from the Association of Pension Companies, more than 75% of savers are over the age of 40. In the case of transformed funds, the average age of savers is over 40 and, as they can no longer join, this average is increasing, and the number of active savers will gradually decrease.

This poses a risk in terms of long-term time returns in terms of free incremental cashflow of PPP projects, as the majority of savers withdraw their savings, including returns, when they reach the age of 60.

For the projects listed in Table 1, the average project duration is 31 years. PPP projects in the transport sector are inherently burdened with significant risk due to the very long project duration.

The risks are a possible significant drop in demand coupled with a drop in revenues. In the case of the DBFMO model, this risk is mostly borne by the investor.

Another risk is a possible unexpected change in project conditions on the part of the government, which may come regardless of any fixed contractual arrangement of conditions at the outset of the project.

These risks have been perceived recently in many situations in the world where they would have been unthinkable before (uncertain developments in Ukraine, rapid changes in the US presidential administration, persistent threats from Russia, the opportunistic exchange rate of Hungary and Slovakia vis-à-vis other EU countries, uncertainty related to the parliamentary elections in the Czech Republic in autumn 2025.

From the available information, it is not possible to reliably determine the resulting profitability of PPP projects, and thus the break-even point in the discounted time return.

Table 1. Examples of 15 PPP rail projects implemented in the EU (Source: based on the analysis of data in [6])

PROJECT Commissioning	Scope	Share of private capital	Type and duration of the contract	Revenue
VRT Sud Europe Atlantique France 2017	EUR 7.7 billion	49%	DBFMO 50 years	Taking inflation into account
VRT Nimes-Montpellier France 2017	EUR 1.8 billion	66%	DBFM 25 years	Taking inflation into account
VRT Bretagne-Pays de la Loire France 2017	EUR 4 billion	33%	DBFM 25 years	?
VRT Perpignan-Figueras France 2010	EUR 1.2 billion	50%	DBFMO 53 years old	?
VRT JIH Netherlands 2009	EUR 1.2 billion	84%	DBFM 30 years	EUR 105 million r. 2009
VRT HS1 UK 2007	GBP 6,8 billion	100%	DBFMO 30 years	Taking inflation into account
Jelly. Tunnel Liefkenshoek Belgium 2014	EUR 0,9 billion	79%	DBMF 42 years old	EUR 204 million per year
Jelly. Track Diabolo Belgium 2012	EUR 1 billion	75%	DBFM 35 years old	Taking inflation into account
Zel. Track Arlanda Sweden 1999	SEK 10 billion	63%	DBMFO 45 years old	Taking inflation into account
GSM-Rail France 2018	EUR 1.6 billion	86%	DBMFO 15 years	?
IE fleet UK 2017	EUR 5.7 billion	90%	DBMF 28 years old	?
Thameslink fleet UK 2015	EUR 2.1 billion	93%	DBFM 25 years	?
Jelly. Station Vialia Vigo Spain 2021	EUR 168 million	100%	rent 22 years old	Taking inflation into account
Sopot Railway Station Poland 2012	113 miles. PLN	63%	rent 12 years	Taking inflation into account

(continued)

Table 1. *(continued)*

PROJECT Commissioning	Scope	Share of private capital	Type and duration of the contract	Revenue
RhonExpress track France 2010	EUR 120 million	100%	DBFMO 30 years	3.25 per annum, always indexed at 2% Investment grant EUR 40 million

Even the direct project participants do not have sufficient evidence, as the decades-long time horizon of the project allows only speculative free cash flow planning. It can therefore be assumed that existing and potential investors demand and will demand an appropriate risk premium over and above normal margins in PPP projects.

5 Benefits of Involving Pension Funds in PPP Projects

5.1 Return on PPP Projects for Pension Funds

Leaving aside the legislative issues, in purely financial terms, the participation of Pension Funds must ensure a discounted return on the funds invested over a time horizon that takes into account the maturity of the participants' deposits.

Let's do a model example. With an average age of the saver of 40 years, we should consider a required time return of 20 years, i.e. up to 60 years of the saver's age. The initial basis discount should be 2.1%, which is the value we have determined from the average return of the transformed Česká spořitelna Pension Fund.

We consider it highly likely that the fund will enter the PPP project with the capital provided by the pension savings participants. The share of foreign capital from the point of view of the pension fund will therefore be 100%.

A potential investment of CZK 100 million (K) at a discount to the current long-term rate of return of 2.1% (i) with a 20-year (P) payback would therefore have to generate a free cash flow (CF) of a minimum value according to

$$\mathrm{K} = \sum_{\mathrm{n}=1}^{P} \frac{CF_{\mathrm{n}}}{(1+\mathrm{i})^{n}} \tag{1}$$

K— Potential investment

P— Number of years

CF— Free cash flow

i— Discount rate

, which comes out to CZK 6.18 million per year, if we do not consider the possible valorisation of revenues, which is not commonplace in projects. If we take the inflation rate for the last 5 years of 3.19% as a discount (source of CSO data), then the minimum permissible free cash flow would have to be CZK 6.84 million per year.

In the case of a required return of 5%, the free cash flow would have to be CZK 8.05 million per year and in the case of a 10% return, CZK 11.75 million per year. In the case of a 20-year payback, these values do not sound unrealistic, quite the contrary, especially from the point of view of the public administration.

5.2 Involving Pension Funds in PPP Projects Is a Win-Win Strategy

According to the State Fund for Transport Infrastructure, a total of 550 billion investments will be needed in 2025–2030, mainly in high-speed lines (Source: estimate based on the SFDI budget [8]). In addition to subsidy programmes, there is therefore a particular opportunity to use private capital for these investments.

On the one hand, there are 2.5 million savers in pension funds, whose savings are being managed unprofitably by the funds, and on the other hand, there is a massive demand for new massive capital for building public infrastructure in the Czech Republic.

The current national debt amounts to CZK 3.4 trillion and the cost of servicing it is around CZK 60 billion a year after deducting outliers. It is now no longer possible to talk about the Czech Republic's debt being low or insignificant, as the CZK 60 billion is missing from the budget every year and continues to increase the debt. In this situation, any loan that is below the long-term implicit interest rate of 3.5% (the average over the last 25 years) is profitable for the state.

The involvement of the Pension Funds in PPP projects therefore starts at a base value of 3.5% as a matter of course from the state's point of view, without the Pension Funds actually increasing their capital participation in the state debt and without the PPP projects causing an immediate increase in the state debt. Yet the 3.5% appreciation alone not only beats the current returns of conservative strategies, it also beats inflation in the long run.

In addition to the above considerations, it should be added that the payback period we considered as described above was 20 years, but Table 1 shows an average contract duration of 31 years. This means that the project payback is considered in this paper with a significant margin of error rather pessimistically.

The PPP sphere is one that pensions have long favoured, offering, at least in theory, a win-win situation [9].

Therefore, if PPP projects are prepared in a meaningful and professional manner, it can be expected that the valuation of the projects will be significantly higher than 3.5% in real terms. An important result is that there is significant scope for finding the optimal distribution of project benefits between the state and the pension funds in terms of a win-win outcome.

6 The Potential for Pension Funds to Participate in PPP Projects

6.1 Involvement of Transformed Pension Funds in PPP Projects

In the case of transformed pension funds, a time horizon of 20 years is not sustainable, as the participants in these funds, given their age structure, will require a shorter return horizon, which will furthermore shorten each year as the average age of participants increases steadily.

For this reason, the direct involvement of transformed pension funds without additional measures is risky in terms of returns and will certainly always be unprofitable in a few years' time. At the same time, however, it is in the transformed funds that the highest potential for their involvement in PPP projects lies.

The way to involve transformed pension funds in PPP projects in a sustainable way is to use the principle of cash pooling. In fact, the vast majority of group operators of transformed pension funds also operate participatory pension funds, where the age structure is different, the average age of participants is lower and does not increase systematically.

The projects of the transformed pension fund may be gradually taken over by the participant fund over the long term in the form of capital provided by cash pooling, and the transformed fund will share the benefits of the project with the participant fund in a corresponding proportion.

This principle will not impoverish the existing participants of the transformed fund. For the promoters of the participation fund, this principle offers another option for the appreciation of their deposits.

6.2 Involvement of Participating Funds in PPP Projects

Participating funds with a conservative strategy can be directly involved in PPP projects, but their capital share is minority.

For funds with a very bold strategy, PPP projects are not suitable as they do not embody the higher risk associated with a potential premium return in the event of a successful development in equity markets.

However, by far the broadest group is represented by managers in balanced strategies, whose investments focus predominantly on the bond market (70% according to [10] statistics as of end-2024).

At the same time, balanced strategies currently more or less only offer variability in the form of adjusting the ratio of bonds to equities in the portfolio, thus offering participants the opportunity to fine-tune their demand for possibly increased certainty or risk, making balanced strategies closer to a conservative or, conversely, a bold strategy.

The inclusion of participant funds with a balanced strategy in the cash pooling together with the transformed pension funds could bring about a significant revival of the existing product offering and contribute to a higher profitability of the pension company as a whole.

6.3 Potential for Pension Funds to Be Involved in PPP in the Czech Republic

Transformed pension funds and participating pension funds with a conservative strategy should be directly involved in PPP projects. Participating pension funds with a balanced strategy should be involved indirectly, for example through cash pooling within pension companies. Participating pension funds with a bold strategy are not suitable for involvement in PPP projects.

Let us summarise these conclusions, including the capital potential, in the table below. From the available data, let us consider the structure of a conservative strategy of 7%, a balanced strategy of 70% and a dynamic strategy of 23% (Table 2).

Table 2. Potential involvement of pension funds in PPP projects in the Czech Republic (Source: prepared based on data from [11])

Fund type	Deposits of savers	Possibility of connection
Transformed	CZK 377 billion	Direct to 20 years horizon
Participatory - conservative	CZK 15 billion	Direct unlimited
Participatory - balanced	CZK 150 billion	Indirect via cashpooling
Participatory - dynamic	CZK 50 billion	Not suitable for wiring

If we consider that the funds will participate in the transport PPP projects with 10% of the deposits, then the above table shows a potential for the Czech Republic of over CZK 50 billion, which could be used in this way, provided that the legislative restrictions that would prevent the entry of pension funds into PPP projects could be removed.

6.4 Analysis of the Development of the Financial System

The quality of institutions affects a wide range of economic characteristics. Higher institutional quality contributes positively to the overall level of wealth, total pension funds to GDP, capital openness and reduces the variability of inflation. Of course, different indicators of institutional quality are correlated. In contrast, legal tradition or openness in foreign trade are relatively independent of the other variables.

For comparison, this paper uses the FDI (Financial Development Index), a comprehensive indicator that aggregates various aspects of financial system development. It shows how developed financial institutions and markets are in terms of their depth, efficiency and accessibility The index is compiled for 183 countries at an annual frequency from 1980 to 2016. The final index is a weighted average of the sub-indices, where the weights are determined by principal component analysis. Attention is paid to both missing data and outlier observations. More information on the construction of the index can be found in Svirydzenko (2016) [12].

7 Conclusions

Currently, the Czech Republic is urgently looking for new optimal sources of financing for transport infrastructure. A number of options have been put forward, including the option of involving the capital of pension funds in PPP projects. In this area, there is even an opinion that without PPP projects, transport infrastructure cannot be built efficiently. There are also opinions that PPP projects in the field of rail transport seem to be the most promising.

However, no in-depth study has yet been produced to address the potential use of pension fund capital in a PPP project. In our article, we have relied mainly on foreign data from the field of PPP projects in rail transport. The current annual budget of the State Fund for Transport Infrastructure consists of approximately CZK 50 billion of investments in the field of rail transport.

The possible involvement of pension funds' capital then has the potential to cover this budget for 1 full year and a theoretical cumulative potential of up to 10 years. This is a unique opportunity that should not be missed by the Czech Republic, on the one hand, and by pension funds and their participants, on the other.

The article can be understood as a starting summary providing initial information for further broader research into the ways to make pension funds' revenues more efficient through their involvement in PPP projects. The article presents the current state of play and shows, based on specific experiences from abroad, that PPP projects could be one of the key paths for the future of Pension Savings in the Czech Republic.

Acknowledgements. We thank the Czech Technical University in Prague, Faculty of Transportation Sciences for the contribution to the conference fee.

References

1. Czech Statistical: Office (2000-2024). https://csu.gov.cz
2. Annual reports of pension fund Česká spořitelna - penzijní společnost, a.s. (2000-2024). https://www.csps.cz
3. OECD: (2018). https://www.oecd.org/en/data.html
4. Ministry of Finance of the Czech Republic: (2024). https://www.mfcr.cz
5. Ministry of Transport of Czech Republic: (2024). https://md.gov.cz
6. Janeba, T., Čoček, T., Kavka, M.: PPP na železnici: Czech Infrastructure Association (Czech Republic) (2022)
7. Šíp, E.: PPP Potential in the Czech Infrastructural Transportation Projects: Stavebnictví ISSN: 1802–2030 (2007)
8. Státní fond dopravní infrastruktury (Czech Republic) (2024). https://sfdi.gov.cz
9. Haurant, S.: Building together, Public-private Partnerships (2024). https://www.europeanpensions.net
10. Office of the Czech Fiscal Council (2025). https://www.rozpoctovarada.cz
11. Czech Nation Bank (2025). https://www.cnb.cz
12. Heidu, R.F.: Pension Systems in Legal and Economic Contexts: Dissertation, Brno (2018)

Factors Shaping the Supply of Traditional Fuels and Their Impact on Price, Analysis of Selected Cases

Piotr Kotylak(✉)

Faculty of Transport, Warsaw University of Technology, Warsaw, Poland
piotr.kotylak@pw.edu.pl

Abstract. The article analyses the main factors shaping the supply of traditional liquid fuels in Europe and their impact on price dynamics. It leveraged the IMRaD framework, combining literature and sectoral data review with scenario analysis based on recent market disruptions such as OPEC + decisions, refinery failures, cyberattacks, and geopolitical conflicts. The results indicate that supply is limited by the availability of raw material, processing capacity, logistics, regulations, and sudden shocks, which quickly translate into retail prices. A recurring pattern of price asymmetry has been identified, i.e. rapid price increases with slower declines. The proposals highlight the importance of diversifying supplies, maintaining strategic reserves and regulatory oversight to increase market resilience. The analysis provides practical guidance for policymakers, energy regulators, and fuel market operators.

Keywords: fuel supply · oil market · refining capacities · supply shocks · price asymmetry

1 Introduction

The stability of liquid fuel supply remains a critical determinant of economic security and social welfare. Despite the global transition toward alternative energy sources, gasoline, diesel, and other petroleum products still dominate transport and industry, making their availability and pricing a direct factor in inflation, competitiveness, and household expenditure. This reality justifies systematic investigation into the mechanisms linking supply disruptions to price volatility, as understanding these processes is necessary for both economic policy and operational decision-making.

The global oil market illustrates the vulnerability of energy systems to production shocks. With OPEC + countries controlling around 60% of exports [1], their collective decisions directly shape market conditions. Production cuts announced in October 2022 (2 million b/d) [2] and April 2023 (1.16 million b/d) [3] reducing supply of global demand. Such episodes underline the importance of monitoring producer policy as a determinant of price stability and as a variable that economic actors must incorporate into risk assessments. The European case provides an equally instructive example of structural dependence and its consequences. Prior to the war in Ukraine, the EU imported

E. Macioszek et al. (Eds.): TSTP 2025, LNNS 1769, pp. 210–224, 2026.
https://doi.org/10.1007/978-3-032-14094-4_14

approximately 25% of crude oil and 40% of diesel from Russia [4]. The sudden disruption of these flows following Russia's invasion in 2022 triggered a 40% surge in global oil prices within weeks [5] and contributed to historically high fuel prices across the Union [6]. These developments demonstrate how geopolitical conflict can expose the fragility of supply chains, while the closure or conversion of 26 European refineries since 2009 [7] has diminished reserve capacity, leaving the downstream sector with limited resilience to shocks. Such vulnerabilities justify closer academic inquiry into the long-term sustainability of Europe's fuel supply architecture.

The relevance of these issues extends to the microeconomic level of retail operators. Price asymmetry, in which wholesale surges are transmitted to consumers more rapidly than declines [8], reflects structural inefficiencies that impact both market fairness and consumer trust. At the same time, infrastructure failures such as the Colonial Pipeline cyberattack in 2021 revealed the disruptive power of technical incidents: 11,667 stations affected experienced shortages, while panic buying pushed prices to seven-year highs [9], even though the average price effect was relatively modest (~0.04 USD/gal) [4]. These examples highlight why continuous research into fuel supply security is indispensable, as it supports the design of effective mitigation strategies at both system and operator level.

2 Methodology of the Study

The structure of the study was adopted according to the IMRaD (Introduction, Methods, Results, and Discussion) convention. Based on a review of scientific literature, industry reports and market data, the main categories of factors shaping the supply of traditional liquid fuels have been identified:

- Availability of raw material
- Refinery processing capacity
- Agility of logistics infrastructure
- Impact of biocomponents
- The impact of legal regulations
- Discontinuous risks (supply shocks)

Analysis required a detailed information tied to market events; the source base contains a large representation of industry reports, press releases, official statistics, news articles, and online datasets that provided event-specific data and illustrative examples. These event-centred sources were used alongside peer-reviewed studies to reconstruct timelines, quantify impacts and extract real-world instances for the scenario analysis.

Based on the identified factors, an important and significant scenario analysis of their impact on final fuel prices was carried out. These scenarios are based on real-world market events over the period 2017–2023 and use real-world empirical data on the impact of such events. Depending on the nature of the data, descriptive or comparative methods were used. To illustrate the asymmetric response of prices, the results of econometric studies by other authors were used (e.g. the ECM model for fuel prices). The adopted methodology provides an illustrative view of the historical determinants of fuel supply, considering wide perspective.

3 Analysis of Supply Factors Based on Historical Cases

This part analyses the impact of the identified factors affecting the supply of fuels and their translation into market prices.

3.1 Commodity: Global Oil Supply and the Role of OPEC +

The Importance of Oil Supply

Crude oil is the basic raw material to produce gasoline, diesel, LPG, kerosene, and many other liquid fuels. Its availability and price directly determine the costs of fuel production in refineries. Historical crises of oil supply (e.g. OPEC embargoes in the 1970s, Gulf War in 1990–91) have shown that a reduction in oil supply causes a sharp increase in oil prices, followed by a sharp increase in fuel prices at petrol stations [1]. In the structure of gasoline prices, in many countries the raw material is the largest component, according to the EIA, about 50–60% of the retail price is due to the cost of oil [1]. Therefore, the higher the price of oil (e.g. due to a shortage of supply), the higher the fuel prices, as long as the other elements (margins, taxes) remain constant.

OPEC and Supply Control

For several decades, global oil supply has been significantly affected by the policy of the OPEC cartel (Citizenship of Oil Exporting Countries) and, since 2016, the expanded OPEC + format, which additionally includes, among others, Russia, Kazakhstan, and other countries. The cartel, which controls about 1/3 of the world's oil production and ~ 60% of exports [1], is trying to control the market balance and price level by setting production limits. When OPEC cuts production quotas, supply in the market shrinks, which (given demand) usually pushes prices up. Conversely, an increase in OPEC production may cool the market and lower prices, although in practice the cartel is less likely to initiate such actions unless it fears a permanent loss of market share to non-OPEC producers. In the 2020–2023 period, both periods of supply expansion (e.g. gradual increase in production after the collapse in demand during the pandemic) and supply constraints were recorded. An example of the latter were the aforementioned decisions from the end of 2022 and the beginning of 2023: In October 2022, OPEC + announced a production cut of 2 million barrels per day from November 2022 [2]. The actual drop in supply was estimated at around 1 million b/d (because some countries were extracting less than the previous limits anyway), but this decision caused political tensions and affected expectations of an increase in oil prices [2]. Indeed, after a period of price declines in the summer of 2022 (Brent from ~ $120 in June to ~ $90 in September), prices stabilized in the $90–$100 range in the fall, supported by the OPEC + supply cut [2]. In April 2023, there was an unexpected, voluntary reduction in production by several OPEC + countries (among others. Saudi Arabia 500,000 b/d, Iraq 211,000 b/d, UAE 144,000 b/d, Kuwait 128,000 b/d, Russia extension 500,000 b/d), a total of ~ 1.16 million b/d from May to the end of 2023 [3]. This move surprised the market and oil prices immediately increased by about 6% (Brent from ~ 80 to ~ 85 USD/bbl) on the news of the cuts [10]. The total effect of all OPEC + reductions in force (including the previous 2 million b/d in 2022) reached 3.66 million b/d, i.e. 3.7% of global demand [10]. Analysts had forecast that the move would keep oil prices high ($100 or more) for the rest of 2023 [10], especially as this came at a time of low global inventories and inflation concerns.

These cases show that coordinated reduction in the supply of raw material by producers results in an increase in raw material prices, and this translates into more expensive fuels for end users in subsequent links of the chain (refineries, wholesale, retail). Importantly, the impact of such OPEC decisions can also be felt in Europe, despite the fact that the EU imports oil from various sources, the global price of Brent (which is also a benchmark for supplies to Europe) reacts to the cartel's actions. In the economic literature, OPEC's actions are sometimes modelled as a factor causing negative supply shocks for oil-importing countries, which in turn translates into a decline in economic activity and higher prices for consumers.

Reserve Capacity and Supply Elasticity

An important parameter related to OPEC is the volume of spare production capacity, i.e. the ability to quickly increase production in the event of a sudden shortage. In recent years, reserve capacities have been concentrated mainly in Saudi Arabia and the UAE. When global supply is tight (demand close to capacity), the market becomes sensitive to any disruption and oil prices show high volatility. For example, in 2022, after a decline in oil investments in previous years and OPEC + production cuts, reserve capacities were estimated at relatively low, which meant that, for example, the loss of most of Russia's production could not be quickly compensated. Hence, among others, the jump in prices after the outbreak of the war in Ukraine: markets discounted the scenario of losing Russian oil, knowing that other countries did not have enough surplus capacity to fill this gap in a short time [5].

Geopolitics and Other Raw Material Factors

In addition to OPEC +, oil supply is also affected by geopolitical factors beyond the control of the cartel: sanctions on producers (e.g. Iran, Venezuela), internal conflicts (Libya, Nigeria), and the production policy of other large players (the US shale sector). In 2020–2021, the dynamic growth of production in the US (after the pandemic recession) partially offset OPEC's activities, but in 2022 the growth rate of US production was moderate in the face of infrastructure constraints and capital discipline of companies. As a result, the burden of market stabilization shifted, among others, to strategic reserves. The IEA coordinated the release of oil from member states' reserves twice in 2022 (a total of 240 million barrels, equivalent to more than 1 million b/d for six months), the largest such move in the organization's history [11]. This measure was intended to alleviate the shortage of raw material after the outbreak of the war and limit the rate of price increases. Indeed, the noticeable supply from reserves (especially from the United States) in the second half of 2022 helped stabilize the oil market temporarily. Brent fell from ~ $120 in June to ~ $90 in September 2022 [2]. However, the depletion of reserves raised questions about the durability of this effect; for example, at the beginning of 2023 in the US, the level of SPR reserves fell to the lowest level in 40 years [10], limiting the possibility of further interventions.

Summary, the global supply of oil (determined by production capacity and producer decisions) is the foundation of fuel supply. Its fluctuations are quickly transmitted to fuel prices. For European consumers, it is important to monitor OPEC + activities and the overall health of the oil market, because even if Europe is not directly involved in each shock (e.g. OPEC cuts mainly concern Asia/the US), the price effect is felt universally through uniform global oil prices. In the following sections, we will move on to the factors that determine the local supply of fuel products, i.e. refineries and logistics, as they determine whether and how the raw material will translate into the availability of finished fuels at service stations.

3.2 Refinery Capacities and Downstream Capacities

Refineries are the link that converts crude oil into usable liquid fuels. Even with an abundance of oil on the market, restrictions on the processing side may cause shortages of gasoline or diesel. The capacity and flexibility of the refining sector are therefore the second key pillar of fuel supply. The supply of products is influenced by two main aspects: (a) nominal processing capacity (i.e. how many barrels of oil per day can be processed by existing refineries in total) and (b) actual capacity utilisation (how many barrels are processed, depending on demand, profitability and availability of the raw material).

Global Refining Capacity vs Demand

Globally, capacity has been growing for decades, mainly driven by investments in Asia and the Middle East. However, in Europe and North America, the opposite trend has occurred – downsizing of the refining sector. In Europe, approximately 26 refineries have been closed since 2009, representing more than 20% of the previously operating plants [7]. According to Concave data, the number of active refineries in Europe fell from nearly 100 in 2008 to around 70 in 2021 [12]. The reasons for among others this include overcapacity due to slow-growing demand, tighter environmental regulations, and competition from refineries outside Europe. The IEA estimates that by 2030, a further 1–1.5 million barrels per day of processing capacity (i.e. about 10–15% of current capacity) in Europe is at risk if the pressures of the energy transition continue [13]. A reduction in processing capacity means a lower potential supply of fuels and lower safety buffers. The situation at the beginning of 2022 illustrated this well: surprisingly strong demand for diesel in Europe and a simultaneous reduction in imports from Russia were met with a limited ability of European refineries to increase throughput. Refining margins rose sharply in 2022, which signalled that there was a shortage of finished fuels relative to demand, because of throughput constraints, but in the long term, the European downstream sector remains in a shrinking capacity trend [13].

Power Utilization Rate and Flexibility

The refinery's ability to respond to changes in demand depends on spare capacity. When the sector is operating close to its maximum capacity, any unplanned shutdown of a refinery can trigger a shortage of product in the market. For example, in July 2017, there was a fire and power failure at Europe's largest refinery, Shell Pernis in Rotterdam (404 thousand b/d). Shell had to stop production and loading of products in an emergency [14]. As Pernis produces a significant percentage of fuels for the ARA (Amsterdam Rotterdam Antwerp) market, the effects were felt immediately: wholesale prices for diesel and gasoline in NWE (Northwest Europe) jumped to their highest levels in several months [15], and refining margins in the region increased further [14]. In other words, the failure of one large refinery caused a local supply shock. In the absence of excess capacity, other plants could not easily cover the gap. This impact was temporary (the Pernis refinery resumed operations after several days), but during this period, station operators had to consider increases in wholesale prices, and some importers imported additional fuels from outside the region, which affected tanker freights [16].

Investments and Modernizations vs Closures

In Europe, there are diverse strategies of companies. Instead of closing refineries, some companies are modernizing them, for example, switching to the production of biofuels (so-called biorefineries). Examples include the ENI refinery in Venice and the Total refinery in La Mède, which have been converted to process vegetable oils. These transformations reduce the supply of classic fuels (because the raw material is no longer oil, and capacities are often decreasing), but they are a response to climate policy. At the same time, however, some EU refineries continue to invest in increasing unit efficiency (improving fuel yields per barrel of oil). Overall, however, the International Energy Agency predicts stagnation or decline in processing in Europe by the end of 2020, in contrast to increases in Asia [13]. A decline in demand for fuels in the long term (increased electromobility, efficiency) may result in overcapacity, but in the short term (until 2030) there is a risk that the pace of closing refineries in Europe will exceed the decline in demand, which would lead to a tighter product market.

Impact on Prices

In conditions of limited processing capacity, the market for petroleum products may "split" from the oil market. This means a situation in which the price of oil may be stable or even decreasing, and the prices of finished fuels are rising due to the shortage of their production capacity. This is practically observed in the aforementioned changes in refining margins. They are a measure of the difference between the price of products and the raw material. When margins rise to high levels, it indicates a limited supply of products. For example, in 2022, the gasoline margin in Europe (the difference between the price of gasoline and oil) also increased to tens of dollars per barrel, which translated into expensive fuel at the stations despite some stabilization of oil prices in the summer. On a retail scale, according to reports by the European Commission and EIA/IEA data, Eurosuper 95 gasoline prices in Poland and Germany exceeded their historical highs (~8 PLN/L in Poland, ~ 2.2 €/L in DE), driven by the refining and logistics factor, and not only by the raw material.

Summary: refining capacities and their utilisation are a bottleneck in fuel supply. In Europe, the downward trend in capacity makes the market more vulnerable to local shocks. A failure of one plant or a delay in deliveries can significantly increase local prices. For the security of supply, it is crucial to maintain a certain surplus of capacity (processing reserve) or, alternatively, the possibility of quick import of fuels from other regions. The policy of states and companies should take these aspects into account to avoid a situation where the lack of processing results in shortages at stations. The next section discusses the issues of logistics infrastructure, which is another condition for the physical availability of fuels.

3.3 Logistics Infrastructure: Transport and Storage of Fuels

The Importance of Infrastructure

The production of fuels in refineries alone does not guarantee their availability to consumers. An efficient supply chain is necessary. The fuel logistics infrastructure consists of pipelines (transporting both crude oil to refineries and finished products to distribution centres), sea transport (tankers transporting oil and products between regions), rail transport and tank trucks (for regional distribution), as well as fuel storage facilities and depots for buffering stocks. Capacity constraints or interruptions in this infrastructure can lead to local shortages, regardless of the global supply balance. In the literature on the subject, the so-called logistic bottlenecks are often indicated as co-determining prices. In the United States, for example, periodic clogging of pipelines or transport channels between coasts results in fuel price differences between regions.

Key Corridors and Points in Europe

In Europe, the following are of critical importance, among others: oil pipelines (Druzhba supplied Russian oil to refineries in Central Europe and Germany, PEOP planned from Adriatic ports), internal networks (e.g. the Druzhba pipeline also has a product line; CEPS – fuel transport system in Central Europe), ARA port network (AmsterdamRotterdamAntwerp) as the main transshipment hub. In addition, maritime transport with finished products: Northern Europe is supplied with diesel supplies from the USA, the Middle East and earlier from Russia. Therefore, the availability of tankers and the patency of routes (e.g., through the Suez Canal) are important.

Infrastructure Risks

The main threats to fuel logistics are technical failures, capacity overload and random or deliberate events.

Disasters and accidents: Hurricanes in the Gulf of Mexico region regularly shut down some U.S. refineries and pipelines for several weeks, affecting global markets (e.g., Hurricane Harvey 2017 caused gasoline prices on the New York Stock Exchange to jump by more than 25%). In Europe, fuel infrastructure is less likely to be affected by weather events (although, for example, floods can damage tracks or roads).

Route blockades: The Suez Canal is a strategic route for the supply of oil from the Middle East to Europe and products from Asia. ~ 9% of the world's maritime oil transport passes through the Canal [17] (approx. 9 million barrels per day in 2023). Blocking it even for a few days can shake up logistics. A famous case: the blockade of the Canal by the container ship Ever Given in March 2021 halted tanker traffic; although it lasted only 6 days, it caused delays in oil deliveries to Europe. The prices of oil and products reacted to the upside for a short time, and after the channel was cleared, they corrected [18]. The event highlighted the fact that Suez is a bottleneck of global importance: about 50 ships a day, cargo worth up to $9 billion a day. Each interruption means serious consequences [17]. Attacks and sabotage: Nowadays, the risk of cyberattacks on infrastructure is growing (the example of the Colonial Pipeline in the US in 2021 is the most serious cyber incident in the fuel sector [9]) and physical sabotage (for example, unexplained explosions, such as the explosions of the Nord Stream gas pipelines in 2022, although gas pipelines have sparked a discussion about protecting energy infrastructure). Colonial

Pipeline as a case study: a ransomware attack forced the operator to shut down a pipeline delivering ~ 2.5 million b/d of fuel (45% of the US East Coast's supply) for 5 days [19]. Despite alternative deliveries by sea and rail, there have been local fuel shortages and a jump in the average U.S. gasoline price above $3/gal (for the first time since 2014) [9]. This incident showed that cybersecurity has become an integral part of the security of fuel supply.

Warehouses and operating stock: Warehouse bases are also part of the infrastructure. Maintaining an appropriate level of commercial fuel stocks has a stabilising effect. They can be released in the event of a short-term shortage, mitigating price spikes. In practice, however, companies often minimize inventories (just Intime strategy) for cost reasons. Therefore, mandatory state stocks play an important role (in the EU, the requirement to have 90 days of net stocks of products/oil [20]). In emergencies, governments can release these reserves. For example, in 2022, Poland released some of its diesel stocks after Russia's attack on Ukraine to stabilize diesel supply and avoid queues at stations. This action, along with coordination within the EU and the IEA, helped to secure the market despite the halt in imports from Russia.

Impact of logistical constraints on prices: If infrastructure cannot keep up with trade flows, regional price differences and local premiums arise. For example, in October–December 2022, just after the announcement of the EU embargo on Russian oil, freight rates for VLCC and Suezmax tankers increased significantly, resulting in additional freight costs of up to several million dollars per transport, or a few dollars premium per barrel of oil delivered to Europe [21]. In addition, interruptions in oil supplies via the Druzhba pipeline (for example, the October 2022 breakdown in Poland and restrictions on transmission to Germany) forced increased imports by sea to ports such as Gdańsk and Rostock, which were logistically more expensive and translated into higher local gas costs [22].

On the product side, in December 2017, the failure of the Forties pipeline (which carries the main part of Brent crude oil) led to a reduction in gas supply in the UK, and wholesale and retail fuel prices jumped. Fuels may have increased by a few pence per liter, with Brent exceeding $65/bbl, [23–25] These cases show that infrastructure constraints, be it freight available or pipelines, can result in regional increases in product and raw material prices, especially when logistics reaches its limits of flexibility.

Summary: logistics infrastructure is a critical factor in ensuring that the fuels produced reach consumers. Its unreliability or overload may locally reduce effective supply and cause price increases regardless of global trends. Operators and regulators should therefore invest in the maintenance and security of infrastructure (including cybersecurity), as well as plan alternative routes and means of transport in the event of a failure of the main corridor.

Biocomponents and RED Regulations: Impact on Traditional Fuels

The growing role of biofuels: In recent years, energy policy has promoted increasing the share of renewable fuels in transport, mainly through the addition of biocomponents to fossil fuels. For the traditional fuel market, this means that part of the demand for transport energy is met by biofuels, which seemingly reduces the demand for petroleum-based fuels. However, from the point of view of the supply of final fuels, refineries and distributors must provide an adequate amount of biocomponents for blending. The RED

II Directive (2018/2001/EU) imposed an obligation on each EU country to achieve a 14% share of energy from renewable sources in transport by 2030 [26]. In practice, this is mainly achieved through liquid biofuels (bioethanol in petrol, biodiesel in diesel oil) and partly electricity in transport.

Impact of biocomponents on the supply of fossil fuels: The introduction of biocomponents means that the effective supply of fossil fuels may be lower than it would otherwise be, as part of the volume at the stations are bio additives. From the refinery's point of view, this means a potential reduction in the production of petroleum fuels (because they will not fit more in the final fuel, if there is a content limit). If the availability of biocomponents were unlimited, this would not be a problem. The mixing plants add the required quantities, and the supply of final fuel is covered. However, in practice, the availability of biofuels can also be limited: it depends, for example, on the harvest of raw materials (rapeseed, cereals), the production capacity of biocomponents, competition for raw materials (e.g. in years of poor harvest, ester prices increase, and their supply may be insufficient).

An example from the market: in 2021, Germany tightened its national target for reducing emissions from fuels (the so-called GHG quota), which de facto forced an increase in the share of biofuels. However, at the same time, the use of palm oil was banned, and "food" biofuels were reduced [26]. This has resulted in a high demand for advanced biocomponents (UCO – post-frying oils, animal fats). Their supply turned out to be insufficient in relation to the new target, which caused tensions in the market. German fuel suppliers have signalled difficulties in complying with the obligation and have had to use a 'rebalancing' system (buying surplus reductions from companies that have exceeded the target). This, in turn, translated into the cost of a penalty or a fee for failure to meet the indicator. As a result, retail fuel prices in Germany could have been a few euro cents higher to cover these costs. Similarly, in Poland, when mandatory blending was temporarily reduced in 2020 (the so-called National Indicative Target was reduced by the government's decision not to overburden Orlen during the pandemic), after restoring a higher level in 2021, domestic companies had to import a lot of biocomponents, which, given the rising prices of these components, could affect the final price of fuel.

RED regulations and pricing: The cost of meeting biofuel requirements is tricked into prices. If the biocomponent is more expensive than the petroleum component, the price of the mixture increases. In 2022, the prices of biocomponents such as FAME and HVO in Europe increased significantly. Platts reported that biodiesel (FAME) prices were often premium against gas low sulfur futures [27], and HVO reached levels several times higher than ultra-low sulfur diesel [26]. As a result, B7 fuel was a few cents per liter more expensive than pure diesel. In situations of biocomponent shortage, a system of substitute penalties or relaxation of regulations is applied, as in the German GHG reduction system [14].

Summary: Regulations on the share of renewable components affect the traditional fuel sector in two ways: (1) by reducing the volume of fossil fuels that can be legally sold (the rest must be bio), and (2) by adding a new cost and supply element – the market for biocomponents. In the long term (energy transition), this may lead to a decrease in demand for petroleum-based fuels, and thus reduce the pressure on their supply and prices. But in the medium transition period, when requirements are growing faster than the supply of biofuels, tensions and costs may arise. To ensure a smooth supply of fuels, it is important that the development of the biocomponents market goes hand in hand with the goals. Otherwise, it is the lack of bio that will become a limiting factor in the supply of final fuels. In the next part, focusing on discontinuous factors, we will present how sudden disruptions (failures, conflicts, attacks) specifically affected supply and prices, referring to case studies.

Discontinuous Risks: Failures, Conflicts, Cyberattacks and Other Supply Shocks The category of discontinuous risks includes events with low frequency of occurrence but high impact (so-called high impact, low probability). In the context of fuel supply, these are mainly: major industrial accidents, armed conflicts and attacks (terrorist, cyber) against infrastructure, and natural disasters. The common feature is the difficulty of predicting and the immediacy of the effect. They often suddenly eliminate a certain part of the supply (raw materials or products) from the market. Their impact on prices can be immediate and intense, but it can also be short-lived if there is a rapid recovery of supply or replacement of shortages.

Industrial Emergencies and Disasters: These most commonly affect refineries and pipelines. The aforementioned fire at the Shell Pernis refinery (2017) is an example of a case where several percent of Europe's processing capacity was lost for several days, which drove up local diesel and gasoline prices [28]. Another example is the explosion at the Philadelphia Energy Solutions refinery (USA 2019) – a plant with a capacity of 335,000 b/d was permanently closed after the explosion. As a result, the northeastern US had to increase fuel imports, and prices in the region increased by several cents/gal relative to the rest of the country [29]. In Europe, it is worth mentioning the incident of 2020: simultaneous technical shutdowns of several large refineries in France and Italy caused temporary problems with France's gasoline supply (imports from the ARA above the norm) [30]. The failure of the transport route may have a similar effect: when the Druzhba pipeline was shut down for several weeks in 2019 due to oil contamination with organochlorine, refineries in Central Europe (Poland, Slovakia and the Czech Republic) had to reduce their throughput and draw on reserves. Wholesale fuel prices then increased by several percent in the region.

Armed conflicts and terrorist attacks: Wars can cut off entire sources of supply. When the civil war broke out in Libya in 2011, exports from this country (about 1.5 million b/d of high-quality oil) stopped. This caused Brent crude oil prices to rise by ~ $20 in a matter of weeks. In Europe, this manifested itself in higher fuel prices, especially since Libyan oil was the raw material to produce gasoline components. Currently, the key conflict is the war in Ukraine, which has not directly destroyed refineries outside Ukraine, but has triggered sanctions and countersanctions [31]. Russia also reduced the supply of finished products to Europe even before the formal embargo, for example, since the summer of 2022 it has limited diesel exports through the Baltic ports [32]. As a result, in autumn 2022, Europe had to pay record premiums for diesel imports from other directions (e.g. India, the Persian Gulf). At that time, diesel prices reached an unprecedented difference from gasoline (in some places, diesel was 1/3 more expensive than gasoline, which is historically rare). This reflected the diesel supply shock resulting from the conflict. Terrorist attacks against infrastructure, on the other hand, are less frequent, but cannot be ruled out. An example was the drone attack on the Abqaiq and Khura is refineries in Saudi Arabia (2019), which temporarily reduced Saudi oil production by as much as 5.7 million b/d (half of their production at the time). Globally, this translated into a dozen or so percent jump in oil prices immediately after the event. Although deliveries were rebuilt within 2 weeks, the incident showed the vulnerability of even the most protected facilities.

Cyberattacks: The Colonial Pipeline incident (2021) in question is a canonical example of the impact of cyberthreats on the physical supply of fuels. Even though the Colonial pipeline system is located in the US, its importance to the East Coast was critical as it provided ~ 45% of the fuels. When a ransomware attack forced the shutdown of pumps, the region experienced a physical shortage (road and rail deliveries were unable to fully replace the pipeline) [19]. In addition to price increases, this resulted in panic buying. Demand has increased rapidly, exacerbating the problem. In some states, short-term sales rationing has been introduced. It can be concluded that in the era of digitization of infrastructure, cybersecurity is becoming an element of security of supply. Europe has not yet seen such a crippling attack on the fuel sector (although there have been attempts to attack the networks of oil companies), but the risk exists.

Natural disasters: In Europe, less visible than in the US (where hurricanes can shut down refineries in Texas and Louisiana). However, for example, an unusually severe winter may immobilize port infrastructure (frozen bays are a problem for sea supplies to the Baltic States) or cause power failures at the refinery (power outages). Climate change may increase the frequency of extreme events in the future.

Dominoes and supply cascades: An important feature of supply shocks is the possibility of a cascading effect. A reduction in one link (e.g. lack of oil due to war) entails a decrease in throughput, which in turn results in a shortage of product, which ultimately results in an increase in prices. If several factors act simultaneously (e.g. war + refinery breakdown + transport strikes), the effects may overlap. For example, in France in autumn 2022, we had a combination: difficulties with diesel imports from Russia (war), failures of several refineries, and strikes by workers in warehouses. This led to a shortage of fuel at many stations in France in October 2022 and forced government intervention. The price of fuel has risen, and the government has temporarily suspended some taxes to relieve drivers. This case shows that the multiplication of shocks is particularly dangerous.

Summary, discontinuous risks are a kind of rare high-impact events of the fuel market. Although rare, they generate sudden price spikes and test the resilience of the system. In the next section, in the form of case studies, we will discuss in detail selected situations from 2017–2023, which will illustrate how the above factors worked in practice and what conclusions can be drawn from their course.

4 Discussion of Results

Fuel supply is governed by the weakest link in the crude–refining–logistics–regulation chain. Crude shortages limit refinery utilization; processing bottlenecks can decouple product prices from crude. Logistic sport and pipeline throughput and freight availability can temporarily dominate both. Regulation (bio-blending mandates and compulsory stocks) adds a further constraint when bio-feedstocks are scarce. Shocks propagate in two stages: an informational phase, when announcements move prices before physical shortages, and a material phase, when prices stay elevated until flows and inventories recover. The process exhibits the "rockets and feathers" asymmetry, reinforced by low short-run demand elasticity, operational frictions, and market concentration. When processing capacity is tight, refinery margins become the primary carrier of price pressure. Rising crack spreads weaken the crude–product link, so downstream metrics explain short-term retail movements better than crude benchmarks. With a shrinking capacity buffer, even moderate maintenance or technical outages can trigger outsized margin gains and a persistent divergence of product prices from crude.

Logistical bottlenecks create regional bonus irrespective of crude trends. Constraints on ports, pipelines, or product-tanker supply raise delivered costs to demand hubs, while re-routing time prolongs price differentials. In such episodes, logistics can dominate both crude and processing until substitute imports arrive and inventories rotate. Infrastructure and cyber incidents convert risk into actual product shortages, producing local stockouts, precautionary buying, and step increases in prices even when inferior alternatives exist. Technical resilience and cyber-security of transport and storage assets therefore have measurable value by shortening re-equilibration and reducing shock amplitude. Compulsory stocks and coordinated international releases act as effective shock absorbers. Policy should calibrate days-of-cover to Europe's structural diesel shortfall and seasonality; firms should prioritize product inventories at bottleneck-exposed sites. OPEC + decisions pass through quickly via expectations and later via margins and logistics to wholesale and retail, with magnitude determined by the regional processing buffer and corridor availability.

Overall, Europe's shrinking refining buffer, tighter regulation, and fragile logistics heighten vulnerability. Resilience rests on redundancy, flexibility, and cyber-secure infrastructure. Short-run fuel price dynamics reflect the interaction of three elements: refinery margins under capacity tightness, the throughput and cost of logistics corridors, and expectations about future supply. Effective risk management requires monitoring these three sets of indicators, not crude prices alone.

5 Limitations of the Study

The analysis has several limitations due to the available data and the scope adopted. Firstly, it covers the years 2017–2023, and the market situation is dynamic, so the conclusions should be treated as valid at the time of writing and requiring ongoing verification. Second, a complete econometric analysis was not used, and the inference was based on descriptive data and literature. Third, the fuel market is linked to other energy markets (gas, electricity), the impact of which has not been analysed in detail. Another limitation concerns the adoption of simplified ceteris paribus scenarios, which do not fully reflect the complex interactions of many factors. The analysis focused on Europe, so not all findings are directly transferable to other regions. In addition, the issues of pricing policy and subsidies, which in some countries strongly shape the relationship between supply and prices, are omitted. As a result, the conclusions are qualitative and indicate clear directions of impact.

6 Conclusions and Practical Recommendations

The analysis shows that the supply of traditional fuels is exposed to numerous risks, and any disruption in this area quickly translates into price jumps. Therefore, station operators should diversify their sources of supply and maintain minimal reserves, while decision-makers should take care of state stocks and coordination of crisis activities within the EU and the IEA. Bottlenecks in the form of limited refining capacity and insufficient logistics capacity further exacerbate the impact of shocks, highlighting the need for investment in infrastructure and mechanisms to maintain excess production capacity. Regulators should monitor the condition of the refining sector and conduct regular stress tests.

Price asymmetry, which weakens consumer confidence, remains a significant challenge. Operators, while willing to slowly reduce prices to improve margins, should apply a more transparent pricing policy, while regulators should develop information tools, such as retail price comparison apps, and strengthen antitrust supervision where necessary. Geopolitics has become one of the most important factors shaping the market, which requires strategic resilience through the diversification of raw material imports and the development of alternative fuels such as biofuels, synthetic fuels, hydrogen or electromobility. Station operators must prepare for the energy transition by expanding their offerings and reducing their reliance on oil market fluctuations.

A quick and efficient response to sudden crises is equally important. Case studies have shown that clear communication between market participants and administrations makes it easier to manage the situation, so industry-specific emergency protocols and regular practice of disruption scenarios are needed.

Summary, although traditional fuels will remain in use for years to come, their market will be subject to significant tensions resulting from geopolitical and transformative factors. Proactive supply management through diversification, reserve maintenance, infrastructure investments, and effective regulation is essential to minimise price shocks and ensure energy security. At the same time, operators should strengthen operational resilience and build credibility with customers to maintain competitive advantages even in volatile conditions.

References

1. Ang, C.: Explainer: what drives gasoline prices. World Economic Forum, https://www.weforum.org/stories/2022/06/explainer-what-drives-gasoline-prices/ (2022)
2. Ghaddar, A., Lawler, A., Edwards, R.: OPEC+ agrees deep oil production cuts, Biden calls it shortsighted. Reuters, https://www.reuters.com/business/energy/opec-heads-deep-supply-cuts-clash-with-us-2022-10-04 (2022)
3. El Dahan, M., Rasheed, A.: OPEC+ announces surprise oil output cuts. Reuters, https://www.reuters.com/business/energy/sarabia-other-opec-producers-announce-voluntary-oil-output-cuts-2023-04-02 (2023)
4. Niccum, J.: Cyberattack on colonial pipeline affected gas prices far less than initially reported, study finds. KU News, https://news.ku.edu/news/article/2021/12/16/cyberattack-colonial-pipeline-affected-gas-prices-far-less-initially-reported-study-finds (2021)

5. Adolfsen, J.F., Kuik, F., Lis, E.M., Schuler, T.: The impact of the war in Ukraine on euro area energy markets. European Central Bank, https://www.ecb.europa.eu/press/economic-bulletin/focus/2022/html/ecb.ebbox202204_01~68ef3c3dc6.en.html (2022)
6. Council of the European Union: impact of Russia's invasion of Ukraine on the markets: EU response. Consilium, https://www.consilium.europa.eu/en/policies/eu-response-russia-military-aggression-against-ukraine-archive/impact-of-russia-s-invasion-of-ukraine-on-the-markets-eu-response/ (2024)
7. Fuels Europe: refinery closures in Europe. Fuels Europe, https://www.fuelseurope.eu/uploads/files/modules/documents/file/1688471212_t5jVec6rmH5nDGJcVke0i3SdpRp7S5XB3fRNXXcf.pdf (2023)
8. Tsvetanov, T., Slaria, S.: The effect of the colonial pipeline shutdown on gasoline prices. Economics Letters 209, 110122, https://www.sciencedirect.com/science/article/abs/pii/S0165176521003992(2021)
9. Kelly, S.: U.S. gasoline prices climb even as colonial reopens after hack. Reuters, https://www.reuters.com/business/energy/us-gasoline-prices-climb-even-colonial-reopens-after-hack-2021-05-17/ (2021)
10. Xu, M., Lee, J., Ghaddar, A.: OPEC+ cuts put $100/bbl oil back in sight. Reuters, https://www.reuters.com/business/energy/opec-cuts-tighten-mideast-crude-supply-asia-europe-cut-refiners-profits-2023-04-03 (2023)
11. International Energy Agency: IEA confirms member country contributions to second collective action to release oil stocks in response to Russia's invasion of Ukraine. IEA, https://www.iea.org/news/iea-confirms-member-country-contributions-to-second-collective-action-to-release-oil-stocks-in-response-to-russia-s-invasion-of-ukraine (2022)
12. Grover, N.: More than 20% of global oil refining capacity at risk, analysis finds. Reuters, https://www.reuters.com/markets/commodities/more-than-20-global-oil-refining-capacity-risk-analysis-finds-2024-03-28 (2024)
13. Norways, K.: Europe's refining sector braces for major downsizing as margins stall. S&P Global Commodity Insights, https://www.spglobal.com/commodity-insights/en/news-research/latest-news/crude-oil/071824-europes-refining-sector-braces-for-major-downsizing-as-margins-stall (2024)
14. Bousso, R.: Shell suspends Pernis refinery loadings after fire. Reuters, https://www.reuters.com/article/markets/currencies/shell-suspends-pernis-refinery-loadings-after-fire-idUSKBN1AF0UR (2017)
15. Sheppard, D.: Diesel prices in Europe rise after Rotterdam refinery blaze. Financial Times, https://www.ft.com/content/63252af6-75e3-11e7-90c0-90a9d1bc9691 (2017)
16. Angell, M.: Tankers await impact of shell refinery fire. TradeWinds, https://www.tradewindsnews.com/tankers/tankers-await-impact-of-shell-refinery-fire/2-1-136227 (2017)
17. Helwa, R., Al-Riffai, P.: A lifeline under threat: why the Suez Canal's security matters for the world. Atlantic Council, https://www.atlanticcouncil.org/in-depth-research-reports/issue-brief/a-lifeline-under-threat-why-the-suez-canals-security-matters-for-the-world/ (2025)
18. Sheldrick, A., Ghaddar A..: Oil prices slid as Suez Canal reopened to traffic. Reuters, https://www.reuters.com/business/energy/oil-prices-slip-focus-switches-suez-canal-blockage-opec-supply-policy-2021-03-30/ (2021)
19. Menn, J., Kelly, S.: Colonial Pipeline slowly restarts as Southeast U.S. scrambles for fuel. Reuters, https://www.reuters.com/business/energy/top-us-fuel-pipeline-edges-toward-reopening-gasoline-shortages-worsen-2021-05-12 (2021)
20. Council of the European Union: Council Directive 2009/119/EC of 14 September 2009 imposing an obligation on Member States to maintain minimum stocks of crude oil and/or petroleum products. Official Journal of the European Union, L 265. https://eur-lex.europa.eu/LexUriServ/LexUriServ.do?uri=OJ:L:2009:265:0009:0023:EN:PDF (2009)

21. Council of the European Union: Russian oil: EU agrees on level of price cap. Consilium, https://www.consilium.europa.eu/en/press/press-releases/2022/12/03/russian-oil-eu-agrees-on-level-of-price-cap/ (2022)
22. Strzelecki, M., Murray, M.: Druzhba pipeline leak reduces Russian oil flows to Germany. Reuters, https://www.reuters.com/business/energy/polish-pipeline-operator-pern-says-leak-detected-druzba-oil-pipeline-2022-10-12 (2022)
23. Chestnei, N.: British energy prices spike on major North Sea pipeline outage. Reuters, https://www.reuters.com/article/business/energy/british-energy-prices-spike-on-major-north-sea-pipeline-outage-idUSL8N1OB4W2 (2017)
24. Vaughan, A.: UK fuel prices could rise by 3p a litre after North Sea pipeline closure. The Guardian, https://www.theguardian.com/business/2017/dec/12/oil-price-hits-30-month-high-north-sea-ineos-pipeline-closure (2017)
25. Reuters: UK North Sea Forties oil pipeline pumps at half capacity. Reuters, https://www.reuters.com/article/us-britain-oil-forties-idUSKBN1EL0VJ (2017)
26. Appunn, K.: CO2 reduction and biofuels in Germany's transport sector - implementing the RED II directive. Clean Energy Wire, https://www.cleanenergywire.org/factsheets/co2-reduction-and-biofuels-germanys-transport-sector-implementing-red-ii-directive (2021)
27. Platts: Platts assesses biodiesel price differentials often a substantial premium to ICE lowsulfur gasoil futures. S&P Global Commodity Insights, https://press.spglobal.com/2020-11-10-S-P-Global-Platts-Launches-First-Hydrotreated-Vegetable-Oil-HVO-values-in-Europe (2020)
28. Renshaw, J.R.: After fire, Philadelphia Energy Solutions to permanently shut oil refinery. Reuters, https://www.reuters.com/article/after-fire-philadelphia-energy-solutions-to-permanently-shut-oil-refinery-idUSKCN1TR09H (2019)
29. Bousso, R., Wight, A.J.: European refiners struggle with outages as maintenance peaks. Reuters, https://www.reuters.com/article/us-europe-refineries-maintenance-idUSKBN20W2F1 (2020)
30. Zhdannikov, D., Yagova, O.: Russian oil flow contamination roils Europe's refiners. Reuters, https://www.reuters.com/article/markets/currencies/russian-oil-flow-contamination-roils-europes-refiners-idUSKCN1S01TT (2019)
31. U.S. Energy Information Administration: EU's ban on diesel fuel from Russia shifts trade patterns. EIA, https://www.eia.gov/todayinenergy/detail.php?id=55739 (2023)
32. Juknevičienė, R.: EU imports of Russian fossil fuels in third year of invasion surpass financial aid sent to Ukraine. Centre for Research on Energy and Clean Air (CREA), https://energyandcleanair.org/publication/eu-imports-of-russian-fossil-fuels-in-third-year-of-invasion-surpass-financial-aid-sent-to-ukraine (2025)

Preliminary Diagnosis of the Impact Factors of the COVID-19 Pandemic on Passenger Rail Transport Based on a Short Survey

Katarzyna Markowska(✉)

Department of Transport Systems, Traffic Engineering, and Logistics, Silesian University of Technology, 40-019, Katowice, Poland
katarzyna.markowska@polsl.pl

Abstract. This study determines the impact of the coronavirus disease 2019 (COVID-19) pandemic on railway transport in Poland, focusing on changes in passenger travel, perceived safety, and post-pandemic travelling. Based on a nationwide survey, the findings reveal a sharp decline in rail travel during the pandemic, driven by safety concerns, reduced seat availability, and a growing shift toward private vehicles. Despite these disruptions, a significant majority of respondents indicated a willingness to return to rail services after restrictions were lifted. The analysis further highlights how demographic factors and perceived conditions influenced decisions to suspend train use. Although temporary shifts in travel modes were observed, rail transport remained essential for work-related and regional journeys. These findings underscore the resilience of Poland's railway system and provide actionable insights for shaping future mobility policies that prioritize public confidence, flexibility, and pandemic preparedness.

Keywords: Railway transport · Travel behavior · Return to rail · Logistic regression · Passenger perception

1 Introduction

The outbreak of COVID-19 in late 2019 marked a significant turning point in global mobility, affecting nearly every aspect of daily life. In Poland, the first confirmed case was recorded on March 4, 2020 [1]. In response, the Polish government implemented a series of containment measures, including lockdowns, school closures, mandatory face masks, limits on gatherings, and the temporary shutdown of public institutions such as universities and kindergartens [2–5]. These measures, while necessary to control the virus's spread, caused a dramatic halt to public mobility and changed how citizens interacted with public transportation systems.

Public transit fell by over 80% early in the pandemic [6], a trend seen in other parts of Europe and surrounding areas. The railway sector, which is vital for Poland's regional and intercity connections, was greatly affected. Passengers decreased sharply, operations were disrupted, and confidence in the safety of shared mobility was deeply tested. Similar

E. Macioszek et al. (Eds.): TSTP 2025, LNNS 1769, pp. 225–243, 2026.
https://doi.org/10.1007/978-3-032-14094-4_15

issues were observed worldwide, with New York residents limiting movement to within 100 feet during the early peak [7], and researchers in China showing how transportation trajectories influenced the spread of the virus geographically [8–10].

These troubles created not only logistical challenges but also larger socio-economic risks. The European Commission recognized the crucial role of railways in achieving sustainable mobility goals and stressed the importance of protecting this sector's resilience [11, 12]. However, addressing passenger fears and ensuring proper hygiene and reliable service became as important as maintaining operational continuity. This study, therefore, aims to understand how the pandemic impacted Poland's railway system by examining travel behavior, safety perceptions, and recovery trends through a nationwide case study.

The remainder of the paper is structured as follows: Sect. 2 presents the literature review, Sect. 3 provides scenarios, cases, and railway history, Sect. 4 discusses findings and methodology, socio-demography, and questionnaire. Results and analysis are discussed in Sect. 5, and Sect. 6 contains conclusions with key implications and limitations.

2 Literature Review

The COVID-19 pandemic delivered a deep economic and social shock across the European Union and globally, with the magnitude of impact varying depending on region and duration [13, 14]. Among the sectors most affected was transportation, particularly rail, which is critical for both human mobility and freight logistics. Disruptions were widely observed across supply chains, freight operators, and public transport networks [12, 15, 16]. Figure 1 clarifies the outbreak of COVID-19 and increasing of the cases in Poland [1], prompting the government to enact sweeping restrictions that included the closure of schools, universities, nurseries, and cultural institutions [2–5]. Mandatory face masks, social distancing rules, and limitations on public gatherings were also implemented to curb the spread of the virus [2, 17].

As a result, public transport usage in Poland plummeted by over 80% during the first wave of the pandemic [6], a decline echoed across global urban transit systems [3, 13, 18]. Notably, long-distance passenger rail, metro, tram, and light rail networks all reported record-low ridership. New entrants in the railway sector were forced to suspend operations, while long-established companies recorded severe losses [19]. In this context, the sustainability and resilience of European rail infrastructure, critical for economic connectivity and social cohesion, became central concerns for policymakers [11, 12].

Numerous studies have examined the impact of COVID-19 on travel mobility and mode preferences, focusing on variations across socio-demographic groups [20, 21]. Research conducted in New York revealed that nearly half of residents remained within a 100-foot radius of their home at the pandemic's peak, illustrating the collapse of personal mobility [7]. Similar findings were reported in China, where public mobility patterns were shown to influence viral transmission [8–10].

In both global and local contexts, researchers noted that safety perceptions have become decisive factors in shaping public transport usage. Several European studies have explored these behavioral dimensions in greater depth. Molin and Kroesen [3], for example, demonstrated that traveler confidence was closely tied to policy measures such as mandatory mask-wearing and enhanced sanitation protocols. Meanwhile, Almlöf et al. [20] found significant disparities in public transport use in Stockholm, with lower-income individuals continuing to rely on transit more frequently due to necessity rather than choice.

De Haas et al. [6] examined post-lockdown behavior in the Netherlands and found that remote work and reduced commuting persisted beyond the crisis phase, suggesting a potential long-term transformation in travel demand. Finally, from 2020 to 2023, research continued to examine the socio-economic effects of COVID-19, with particular attention to mobility inequality, recovery strategies, and behavioral adaptation across different transport sectors [22–26]. Ensuring safety, accessibility, and trust in rail travel remains central to post-pandemic recovery, particularly in contexts where public transport is essential to equitable urban mobility [18].

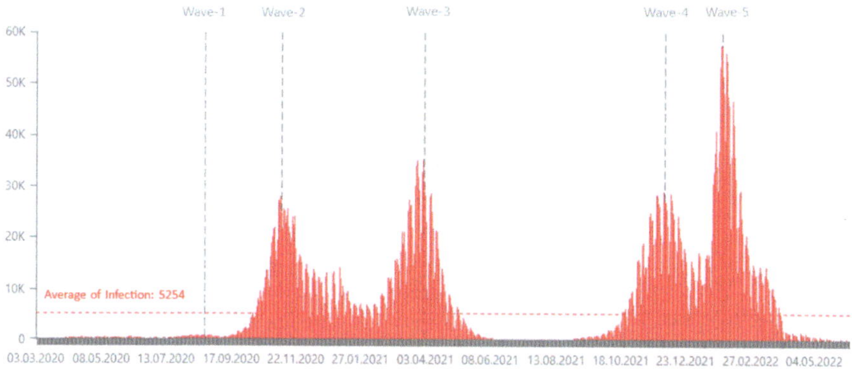

Fig. 1. Confirmed SARS-CoV-2 infections in Poland (2020–2022). Source: Own study based on the data presented in [1]

3 Case Study and Research Question

This study investigates the impact of the COVID-19 pandemic on the functioning of Poland's railway transport system. As a critical component of national mobility and economic connectivity, the rail sector faced abrupt disruptions, rapid policy adaptations, and evolving user behavior. Through an integrated analysis of transport usage data, government responses, and passenger perceptions, this case study provides a grounded perspective on the resilience and recovery trajectory of rail transport in a Central European context.

To guide the analysis, the following hypotheses are proposed:

Hypothesis 1: There is a statistically significant relationship between the type of railway carrier used and the travel frequency of passengers during the COVID-19 pandemic.

Hypothesis 2: The level of trust in rail safety measures (cleanliness, disinfection availability, etc.) is a strong predictor of post-pandemic travel intentions.

3.1 Government Response and Impact on Rail Transport

In response to rising infection rates, the Polish authorities imposed a range of public health measures nationwide lockdowns, closure of educational and commercial institutions, limits on public gatherings, mandatory mask usage, and mass awareness campaigns [1, 2, 5]. These measures, while effective in limiting the virus, significantly disrupted public transport, particularly railways.

A sharp decline in rail transport was observed in 2020, with Q2 ridership falling by 71% compared to the previous year [27]. Figure 2 illustrates how usage closely followed the timing of lockdowns. In Fig. 3, a brief recovery occurred in September 2020 (21.6 million passengers), but numbers dropped again in November due to renewed restrictions. By December, ridership slightly rebounded to 13.3 million. In 2021, recovery gained momentum as restrictions eased, and vaccinations progressed. September marked the highest ridership since the pandemic began [27, 28].

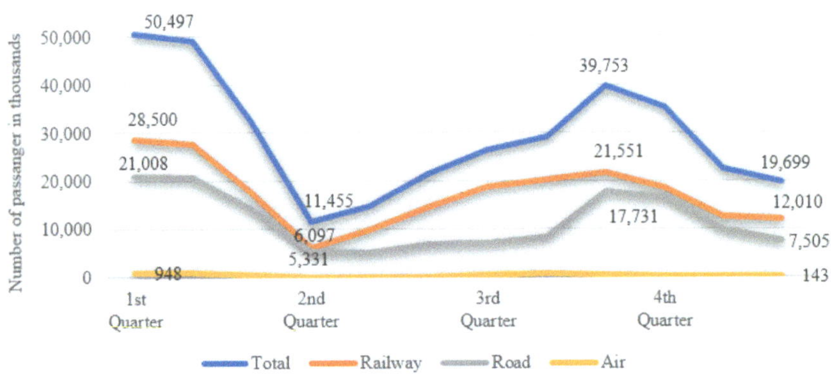

Fig. 2. Passenger Rail Transport in Poland During COVID-19. Source: Own study based on the data presented in [27]

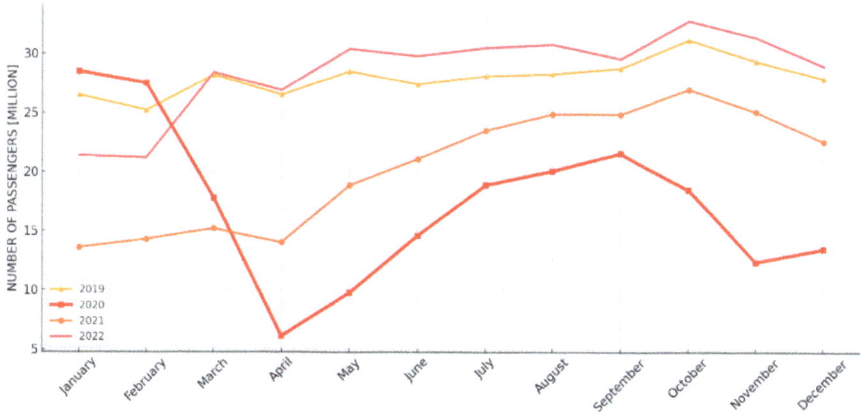

Fig. 3. Monthly Rail Travel in Poland (2019–2022). Source: Own study based on the data presented in [27]

3.2 Strategic Role of Poland in European Rail Corridors

Poland's geographic centrality positions it as a pivotal rail hub in the European and Eurasian logistics networks. It lies on the New Eurasian Land Bridge, connecting China to Western Europe via Kazakhstan, Russia, Belarus, and Poland, a key axis in China's Belt and Road Initiative since its formal launch in 2011 [29]. Furthermore, Poland is a vital component of the EU's Trans-European Transport Network (TEN-T), situated along major corridors such as the Baltic-Adriatic and North Sea-Baltic routes [30].

Investments in national railway infrastructure, primarily by Polskie Koleje Państwowe (PKP Polish Railways), have focused on modernization and capacity enhancement across both freight and passenger segments. However, the pandemic highlighted vulnerabilities even in such strategically important systems. National and international freight routes faced operational delays, while new and incumbent rail operators dealt with logistical bottlenecks and financial strain [13, 18, 19]. The Polish rail database thus illustrates how global health crises can disrupt not only urban mobility but also transnational transport systems crucial to economic resilience.

4 Survey Design and Methodology

4.1 Survey Distribution

To evaluate the impact of the COVID-19 pandemic on railway transportation in Poland, a structured nationwide survey was developed using Google Forms. The questionnaire aimed to gather comprehensive insights into passenger behavior, travel frequency, mode preferences, safety perceptions, and post-pandemic intentions regarding the resumption of train use. The survey was disseminated via social media platforms, mailing lists, and university networks to ensure a broad geographic and demographic reach. Responses were collected over four months and included participants from all 16 voivodeships of Poland, ensuring full national representation (see Fig. 4).

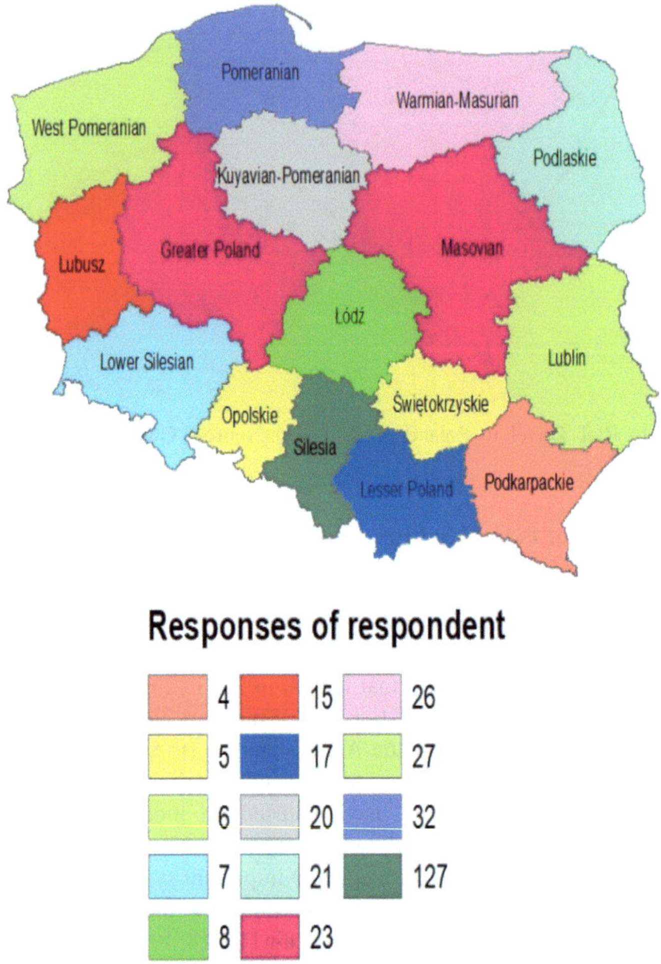

Fig. 4. Geographic Map of (Poland) 16 Voivodeships and Distribution of Survey Responses

To determine the statistical acceptance of the survey sample, Yamane's (1967) and Krejcie & Morgan's (1970) [31, 32] help to validate the margin of error.

$$P = \frac{n}{1 + n(x)^2} \tag{1}$$

where P is the required sample size, n is the total population (30 million), and x is the acceptable margin of error. Using a 95% confidence level and 5% margin of error ($x = 0.05$) Putting the values in Eq. (1):

$$P = \frac{38{,}000{,}000}{1 + 38{,}000{,}000(0.05)^2} \approx 400 \tag{2}$$

From Eqs. 1 and 2, our sample closely approaches this threshold, i.e., 366 responses, and no claim was made regarding a regional or specific area.

4.2 Socio-Demographic Characteristics

A total of 366 Polish residents completed the survey. Of these, 46.7% identified as male and 53.3% as female, with respondents ranging in age from 15 to over 70 years. The age distribution is presented in Table 1. The majority (31.4%) were aged 18–25, while 19.6% were aged 26–40. A small portion (7.4%) was under 18. The estimated mean age group was 36–40 years.

Table 1. Age Distribution of Respondents

Age	Frequency (N)	Percentage (%)
up to 18 years	27	7.37
18 to 25	115	31.42
26 to 40	72	19.60
41 to 50	64	17.48
51 to 60	46	12.56
over 60 years	42	11.45
Mean	26 to 40 years	

4.3 Education Level and Socio-Demographics

Educational attainment among respondents was relatively high. Nearly half (47.3%) had a medium-level education (e.g., high school or vocational training), while 31.4% held higher education degrees. A smaller share included professionals (12.6%) and those with only basic education (8.7%). The education breakdown is shown in Table 2.

The socio-demographic dataset included gender, age, education, and employment status. Additional attributes related to travel behavior and government policy responses (e.g., travel restrictions) were also analyzed. Most respondents reported being employed and relying on train transport for work commutes. As illustrated in Fig. 5, work-related travel was the most frequent purpose, which was severely disrupted during peak COVID-19 waves.

Table 2. Education level of respondents

Category	Frequency (N)	Percentage (%)
Basic	32	08.7
Professional	46	12.6

(continued)

Table 2. (*continued*)

Category	Frequency (N)	Percentage (%)
Medium	173	47.3
Higher	115	31.4

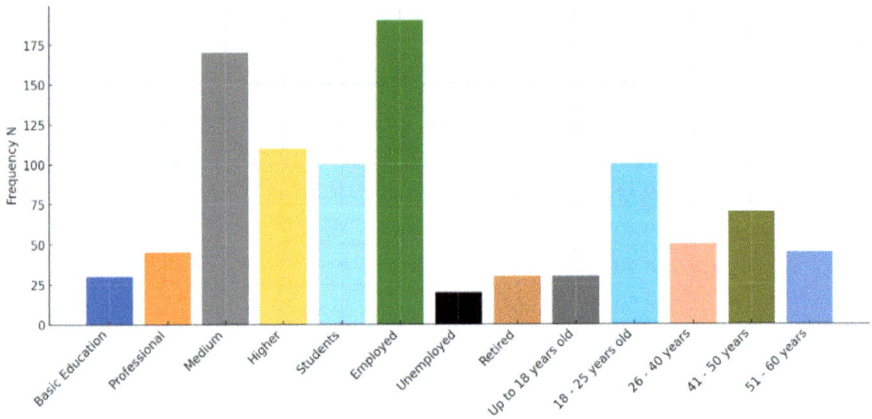

Fig. 5. Socio-Demographic Characteristics of Respondents

4.4 Questionnaire Design and Sections

The survey consisted of five core sections, and question formats included multiple-choice, Likert scale (1 to 5), and binary (Yes/No) options, enabling both qualitative insights and quantitative statistical analysis (such as chi-square tests, logistic regression, etc.).

1. Socio-demographic profile: age, gender, education level, and employment.
2. Rail travel behavior: frequency and purpose of travel before, during, and after COVID-19 restrictions.
3. Transport mode preferences: public versus private transport usage during the pandemic.
4. Perceptions of safety and cleanliness: infrastructure and hygiene ratings.
5. Behavioral shifts: willingness to abandon or resume rail travel post-pandemic.

5 Results and Discussion

This section presents the empirical findings from the survey and highlights key insights into how the COVID-19 pandemic influenced passenger rail usage in Poland. It explores travel frequency, cleanliness and safety perceptions, trip purposes, modal preferences, and the results of statistical modeling.

5.1 Travel Behavior, Cleanliness, and Perceived Safety

The frequency of rail travel showed considerable variation before, during, and after the COVID-19 outbreak. Before the pandemic, daily rail travel was reported by 76 individuals. This figure declined significantly to 29 during the peak of COVID-19 restrictions, indicating a sharp drop in everyday commuting. However, as restrictions eased, daily usage rebounded to 70, demonstrating a strong return to habitual rail use post-pandemic. A similar trend was observed in passengers traveling several times per week. Before the pandemic, 116 respondents indicated this level of use, which dropped to 81 during the outbreak, but increased again to 110 afterward. The most significant behavioral shift was noted among those traveling less than once per month. Only 76 respondents fell into this category before the pandemic, yet this figure rose to 117 during the crisis, a reflection of minimized mobility for health or policy reasons. Post-pandemic, the number of respondents in this category dropped back to 76, mirroring pre-pandemic levels. These changes confirm that both mandatory restrictions and risk perception influenced travel frequency, though many passengers resumed prior behaviors after restrictions were lifted. The pattern is illustrated in Fig. 6, which provides a comparative breakdown of travel habits across the three time frames.

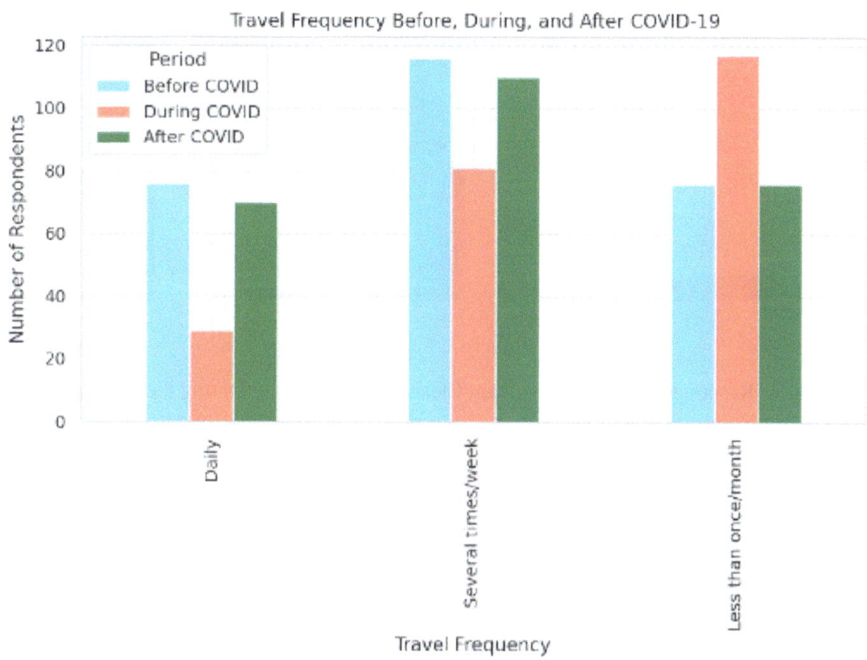

Fig. 6. Frequency of Rail Travel Before, During, and After COVID-19

The survey also sought to assess public perception regarding hygiene and safety on trains during the COVID-19 pandemic. Respondents were asked to rate their perception of rail cleanliness on a scale from 1 (very poor) to 5 (very clean). The results indicated a generally positive assessment. A significant number of respondents (121) rated cleanliness at the highest level, suggesting they perceived rail operators to be maintaining adequate sanitation practices. Another 101 respondents gave a rating of 4, while 96 selected 3, indicating a moderate but acceptable level of cleanliness. Fewer respondents reported lower ratings, suggesting that extreme dissatisfaction with sanitation was not widespread. These findings are depicted in Fig. 7.

Similarly, the perception of travel safety was evaluated using a 1–5 scale. The highest concentration of responses was at level 3 (128 responses), followed closely by level 4 (109 responses), which together comprised over 64% of the total. These results suggest that, although travelers were not overwhelmingly confident, they generally felt moderately safe using rail services during the pandemic. Only a small number rated safety at the lowest levels. This reveals that while there was room for improvement, safety perceptions did not fall below an acceptable threshold for most users. This evaluation is presented in Fig. 7.

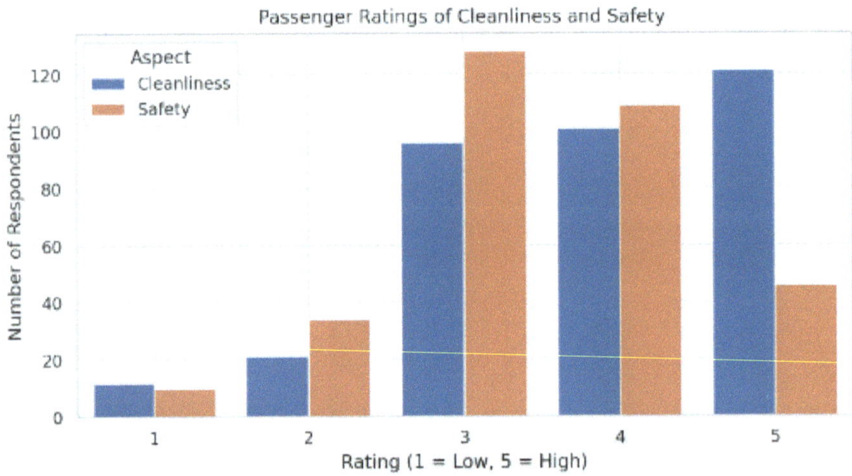

Fig. 7. Respondents' Ratings of Train Cleanliness and Safety During the Pandemic

5.2 Purpose of Rail Travel and Destination Preferences

The assessment revealed multiple purposes behind respondents' rail travel during the COVID-19 pandemic, reflecting both essential and non-essential mobility needs. The most common purpose was commuting to work, reported by 145 individuals, underscoring the central role of railways in supporting the labor force during times of uncertainty. Private and recreational travel accounted for the second-largest group with 134 responses, suggesting that non-essential travel began recovering as restrictions gradually eased.

Business travel was cited by 76 respondents, while university and school commutes were reported by 67 and 30 individuals, respectively. These figures demonstrate that educational and professional obligations continued to shape travel decisions even during periods of limited mobility. These results are visually presented in Fig. 8a, confirming that work and leisure remained the dominant rail travel purposes.

When examining travel distances, most respondents traveled within 50 km, with 39.6% commuting between 21–50 km and 27.0% covering 51–100 km. This reflects a predominance of short- to mid-range travel, typical of urban and suburban commuting patterns. Long-distance travel (above 100 km) was less common (15.3%), likely due to regional restrictions or risk avoidance. Figure 8d highlights this distribution, indicating that rail primarily served short-to-medium corridor needs during the pandemic.

The motivations for using rail transport were also analyzed. As shown in Fig. 8b, price was the most influential factor, cited by 275 respondents, followed by comfort (176) and schedule frequency (141). These preferences emphasize the perception of rail as an economical, convenient, and reliable mode of transport, particularly important during a period of economic and social disruption.

A shift in transport mode preferences was also observed. Figure 8c illustrates that although 200 respondents continued using rail, private cars slightly surpassed trains, with 232 users during the pandemic. This shift was likely driven by individual safety concerns, flexibility, and reduced risk of infection. However, the continued use of rail for essential purposes, especially work and education, suggests that the modal shift toward private cars was situational and temporary.

The convergence of trip purpose, distance, and motivation underscores rail's resilience and adaptability. Despite public health restrictions, the railway system in Poland continued to serve as a critical mobility backbone for both daily commuters and discretionary travelers. The results demonstrate that cost-effectiveness, comfort, and regional accessibility made rail transport a preferred option during and after the pandemic, particularly for short- and medium-distance travel aligned with employment and essential activities.

To determine whether there was a statistically significant association between the purpose of rail travel and the typical travel distance during the COVID-19 pandemic, a Chi-Square Test of Independence was conducted. The two categorical variables Trip Purpose (such as Getting to Work, University, Business Trip, Holiday Travel) and Distance Traveled (0–20 km, 21–50 km, 51–100 km, > 100 km) were cross-tabulated. Statistical dependence was evaluated using the following formula:

$$x^2 = \sum \frac{\left(O_{ij} - E_{ij}\right)^2}{E_{ij}},$$

The test yielded a Chi-square value of $x^2 = 103.70$ with 9 degrees of freedom and a p-value < 0.00001, indicating a highly significant relationship between travel purpose and distance traveled.

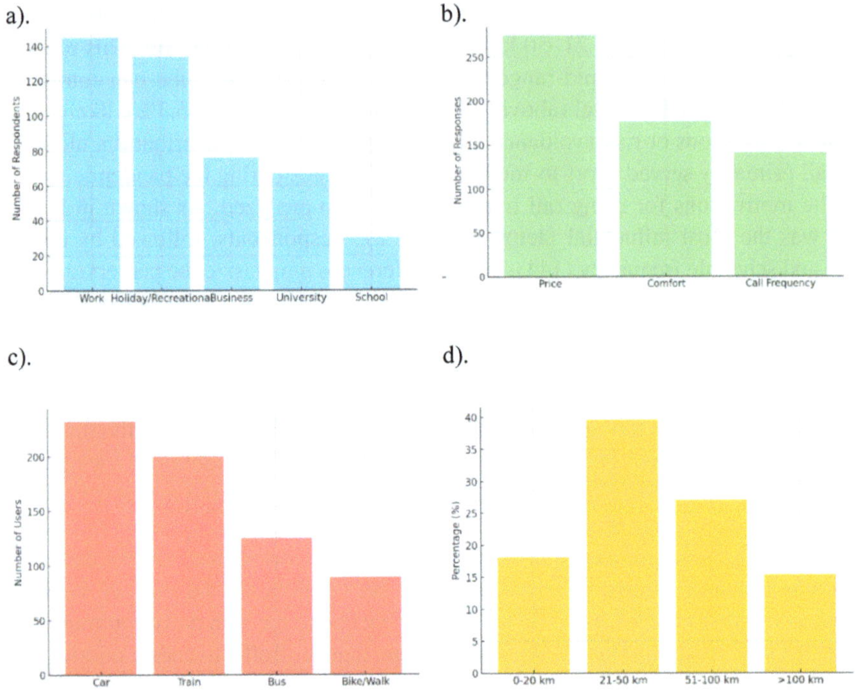

Fig. 8. Rail Travel Purpose, Preferences, and Trip Characteristics a) Primary Purpose of Rail Travel b) Key Motivations for Choosing Rail c) Preferred Modes of Transport d) Distance of Typical Rail Trips

The observed relationships are both statistically and behaviorally meaningful. As shown in Fig. 8, rail travel for work was most concentrated in the 21–50 km range (40 respondents) and secondarily in the 0–20 km range (19 respondents), reflecting typical suburban commuting patterns. Similarly, university-related travel was clustered within 21–50 km (25 respondents), supporting the idea of intercity student movement from residential areas to education hubs (Fig. 9).

In contrast, holiday or leisure travel was most common in the > 100 km category (33 respondents), highlighting the long-distance nature of recreational journeys. Business travel, however, was more evenly spread across all distance categories, suggesting varied professional mobility needs.

This heatmap visualization confirms that each travel purpose exhibits a distinct and logical distance profile, reinforcing the versatility of the Polish railway network in serving short, medium, and long-range needs during the pandemic period.

These findings are consistent with previous research. Almlöf et al. [20] found that essential travel, particularly work and education, continued despite restrictions, often over short to medium distances. Meanwhile, leisure travel was either postponed or shifted to longer distances once restrictions were eased. The significant correlation observed here underlines the adaptive role of rail transport in meeting these evolving mobility demands.

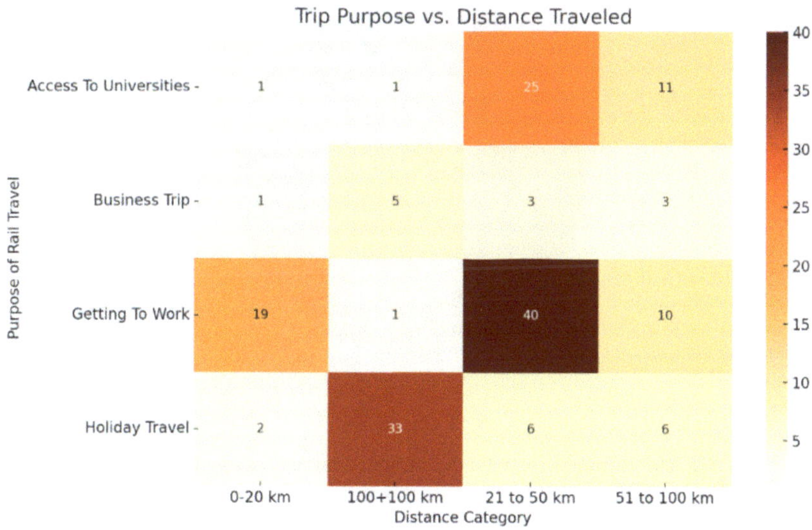

Fig. 9. Relationship Between Travel Purpose and Distance (Chi-Square Test)

5.3 Statistical Modeling and Behavioral Changes During COVID-19

To analyze how user behavior evolved during the COVID-19 pandemic, two binary-response questions were included in the survey. The first asked whether respondents stopped using rail transport at any point during the pandemic. The majority 70.8% answered "No," indicating continued reliance on rail services despite public health concerns, while 29.2% reported discontinuing rail use temporarily. The second question addressed behavior after restrictions were lifted: among those who had stopped traveling by train, 81.3% resumed rail usage, highlighting a strong return rate and suggesting that most disruptions were situational rather than permanent.

These results, shown in Figs. 10 and 11, illustrate the temporary nature of rail abandonment and underscore the resilience and loyalty of Polish rail users during a crisis. They suggest that safety measures and restored public confidence were effective in re-engaging users' post-pandemic.

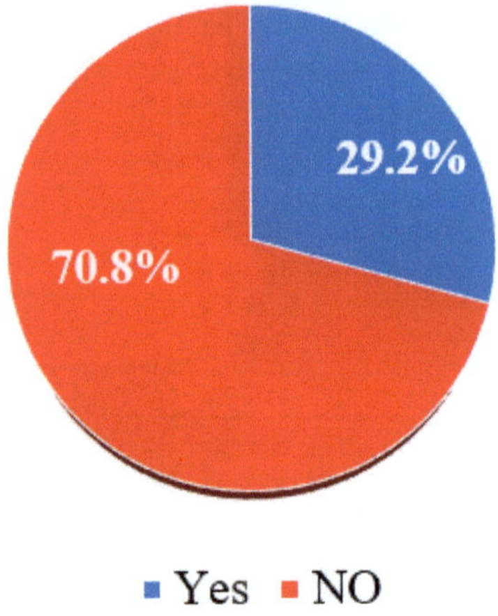

Fig. 10. Respondents who discontinued rail use during COVID-19

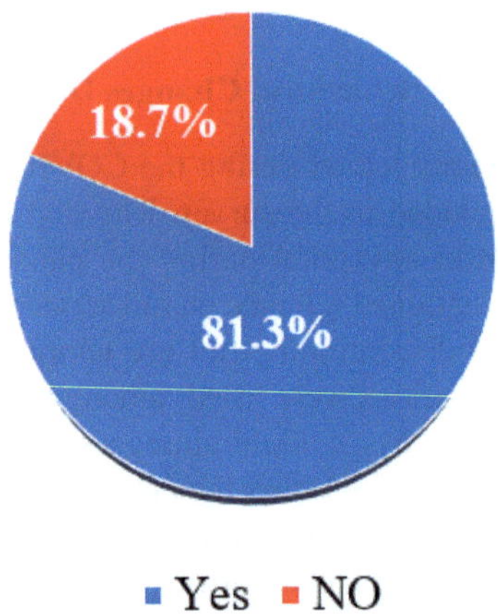

Fig. 11. Traveller who returned to the rail post-restriction

To complement these insights, a logistic regression model was developed to quantify which factors significantly predicted the likelihood of abandoning rail transport during the pandemic. The dependent variable was defined as:

$$Y = \begin{cases} 1, \text{ if respondent gave up rail during COVID-19} \\ 0, \text{ if continued rail use} \end{cases}$$

The general form of the logistic regression model is:

$$\log\left(\frac{p}{1-p}\right) = \beta_0 + \beta_1 X_1 + \beta_2 X_2 \ldots + \beta_k X_k \quad (4)$$

where p is the probability of abandonment, β_k are the coefficients estimated through maximum likelihood, $X_1, \ldots X_k$ is independent variables such as safety perception, age, distance, gender, education.

The model results, presented in Fig. 13, indicate that perceived safety had the strongest influence. Respondents who rated rail travel as clean and safe were significantly less likely to abandon rail, with an odds ratio (OR) of 0.53.

Conversely, age emerged as a strong positive predictor: individuals over 60 years old were 2.36 times more likely to stop using rail during the pandemic, which aligns with research by Almlöf et al. [20] on mobility reductions among vulnerable populations. Additionally, those traveling more than 100 km were more likely to abandon rail, possibly due to increased interregional travel restrictions or fear of long exposures.

Interestingly, gender and education level showed no significant influence, as their odds ratios crossed the null value. This suggests that rail abandonment was more closely tied to structural and health-related concerns than to demographic variables alone.

Finally, the time-series analysis in Fig. 12 illustrates the disruption and gradual recovery of rail usage in Poland. After a sharp drop in early 2020, ridership began to recover steadily through 2021 and 2022, reaching or even exceeding pre-pandemic levels. This U-shaped rebound highlight both the vulnerability and adaptability of the rail sector under crisis conditions.

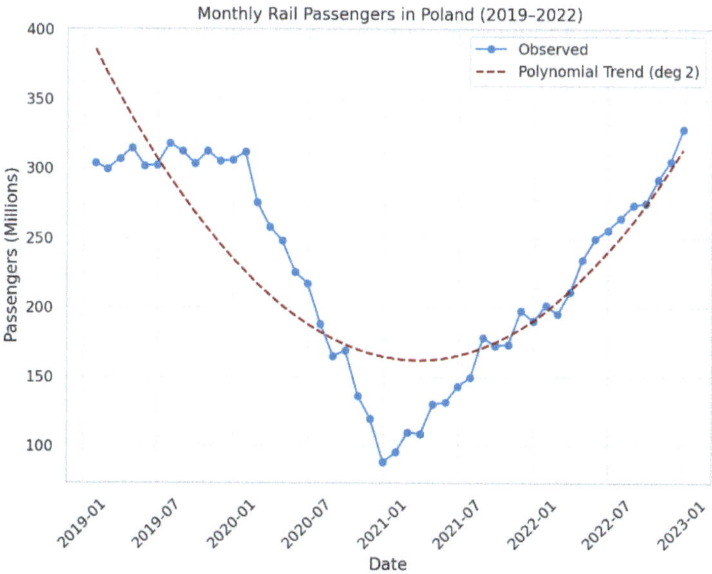

Fig. 12. Monthly Rail Passenger Trends in Poland (2019–2022)

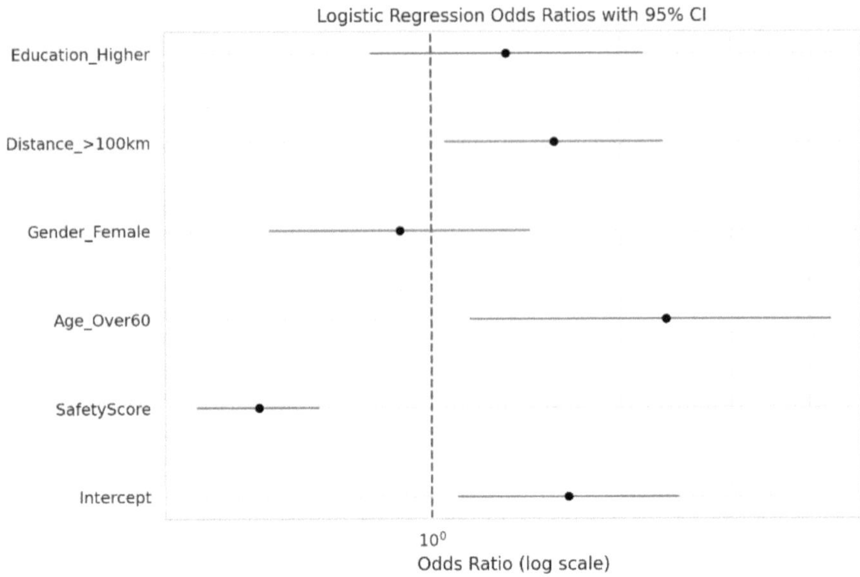

Fig. 13. Logistic Regression Results (Key Predictors of Rail Abandonment)

5.4 Overall Perceived Impact on Rail Transport Functioning

To gauge public sentiment regarding the broader consequences of the COVID-19 pandemic on rail operations, respondents were asked to rate the overall impact on a scale from 1 to 5, where 1 indicated no significant disruption and 5 indicated a very large impact. The most frequently selected rating was 5 (very large impact), chosen by 33% of respondents, followed by rating 4 (27.6%) and 3 (26.2%). In total, more than 60% of participants perceived the impact as either significant or very significant, indicating that the pandemic was widely acknowledged to have affected rail transport beyond routine fluctuations.

This is visually illustrated in Fig. 14, which shows a skew toward higher impact ratings. By contrast, only a small fraction of respondents assigned low ratings (1 or 2), suggesting that few passengers viewed rail operations as unaffected by the crisis. These findings support the broader conclusions drawn throughout this study: while Poland's rail transport sector demonstrated resilience and gradual recovery, the pandemic clearly disrupted its functionality, passenger behavior, and operational dynamics. However, the relatively strong return to rail use, as shown in earlier sections, indicates that these disruptions were substantial but not irreversible, and that strategic planning, health interventions, and user trust played a central role in stabilizing public transport post-crisis.

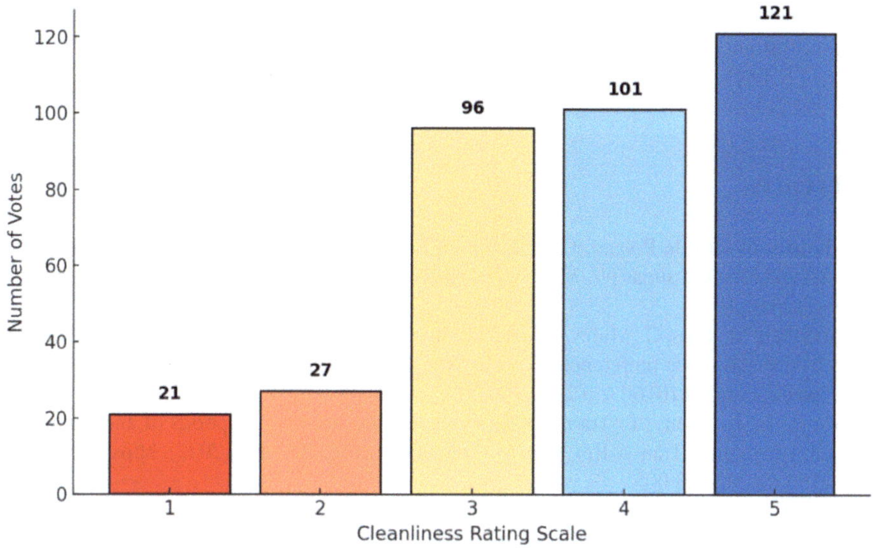

Fig. 14. Respondents' Assessment of the COVID-19 Pandemic's Impact on Rail Transport Functioning (1 = No impact, 5 = Very large impact)

6 Conclusions

This study offered a preliminary assessment of the COVID-19 pandemic's impact on Poland's rail transport system, and the results highlight significant disruptions, including reduced travel, altered travel behaviors, and increased reliance on private vehicles during peak restriction periods. Despite these challenges, the findings underscore the resilience of the railway sector and user adaptability 81% of respondents expressed willingness to resume train travel once restrictions eased, indicating continued trust in rail services. Rail remained essential for commuting and education, particularly within regional distances, and was favored for its affordability, comfort, and accessibility. Positive perceptions of cleanliness and safety reflected the effectiveness of sanitary measures introduced by operators. While the pandemic posed short-term setbacks in usage and revenue, the long-term outlook remains optimistic. This case study offers actionable insights for strengthening the resilience of rail systems through improved health communication, operational preparedness, and user-centered planning, key to fostering sustainable and crisis-ready transport across Poland and the broader European context.

7 Limitations and Future Research

1. Although 366 responses were collected, the sample may not represent the full demographic and regional diversity of Poland, particularly rural or underrepresented areas. Future research should adopt stratified sampling or increase outreach in under-sampled voivodeships to enhance generalizability.
2. Respondents' answers were based on personal perceptions and memory, which may have introduced inaccuracies. Future work should triangulate survey data with real-time behavioral or ticketing data to improve reliability.

3. As a cross-sectional study, the research captured only a single phase of the pandemic. Longitudinal follow-ups and simulation-based models are needed to analyze evolving travel behavior and assess preparedness for future transport disruptions.

References

1. Ministry of Health, Poland, Coronavirus in Poland, current and historical data and charts. https://koronawirusunas.pl/. Ministry of Health, Poland, 19 September 2023. https://koronawirusunas.pl/
2. Abdullah, M., Dias, C., Muley, D., Shahin, Md.: Exploring the impacts of COVID-19 on travel behavior and mode preferences. Transp. Res. Interdisciplinary Perspect. **8**, 100255 (2020). https://doi.org/10.1016/j.trip.2020.100255
3. Molin, E., Kroesen, M.: Train travel in corona time: safety perceptions of and support for policy measures. Transp. Res. Part A: Policy Pract. **158**, 196–209 (2022). https://doi.org/10.1016/j.tra.2022.03.005
4. Paul, T., Ornob, A.B.S., Chakraborty, R., Anwari, N.: Assessment of COVID-19 induced travel pattern changes in Dhaka City. Case Stud. Transp. Policy **9**(4), 1943–1955 (2021). https://doi.org/10.1016/j.cstp.2021.11.003
5. Kwok, K.O., et al.: Community responses during early phase of COVID-19 epidemic, Hong Kong. Emerg. Infect. Dis. **26**(7), 1575–1579 (2020). https://doi.org/10.3201/eid2607.200500
6. De Haas, M., Faber, R., Hamersma, M.: How COVID-19 and the Dutch 'intelligent lockdown' change activities, work and travel behaviour: evidence from longitudinal data in the Netherlands. Transp. Res. Interdisciplinary Perspect. **6**, 100150 (2020). https://doi.org/10.1016/j.trip.2020.100150
7. Warren, M.S., Skillman, S.W.: Mobility changes in response to COVID-19 (2020). https://doi.org/10.48550/ARXIV.2003.14228
8. Chinazzi, M., et al.: The effect of travel restrictions on the spread of the 2019 novel coronavirus (COVID-19) outbreak. Science **368**(6489), 395–400 (2020). https://doi.org/10.1126/science.aba9757
9. Kraemer, M.U.G., et al.: The effect of human mobility and control measures on the COVID-19 epidemic in China. Science **368**(6490), 493–497 (2020). https://doi.org/10.1126/science.abb4218
10. Linka, K., Peirlinck, M., Sahli Costabal, F., Kuhl, E.: Outbreak dynamics of COVID-19 in Europe and the effect of travel restrictions. Comput. Methods Biomech. Biomed. Eng. **23**(11), 710–717 (2020). https://doi.org/10.1080/10255842.2020.1759560
11. Knapcikova, L., Konings, R.: European railway infrastructure: a review. AL **5**(3), 71–77 (2018). https://doi.org/10.22306/al.v5i3.97
12. European Union, Establishing Measures for a Sustainable Rail Market in View of the Covid-19 Pandemic (2020). https://eur-lex.europa.eu/legal-content/en/TXT/?uri=CELEX%3A32020R1429
13. WSP Global, Rail and the Effects of the COVID-19 Pandemic, 2021 (2021). https://www.wsp.com/-/media/Insights/Global/Documents/White-Paper---Rail-and-the-Effects-of-the-COVID19-Pandemic.pdf
14. IRG Rail, Impacts of the COVID-19 crisis and national responses on European railway markets in 2020 (2021). https://irg-rail.eu/download/5/887/IRG-Rail20215Covidpublication2021.pdf
15. Gdynia Maritime University and A. S. Grzelakowski, Global supply chains disruptions as a result of the COVID-19 pandemic crisis and the dynamic world container shipping market recovery. GMiL **2022**(12), 10–20 (2022). https://doi.org/10.33226/1231-2037.2022.11.2

16. Rokicki, T., Bórawski, P., Bełdycka-Bórawska, A., Szeberényi, A., Perkowska, A.: Changes in logistics activities in Poland as a result of the COVID-19 pandemic. Sustainability **14**(16), 10303 (2022). https://doi.org/10.3390/su141610303
17. Jenelius, E., Cebecauer, M.: Impacts of COVID-19 on public transport ridership in Sweden: analysis of ticket validations, sales and passenger counts. Transp. Res. Interdisciplinary Perspect. **8**, 100242 (2020). https://doi.org/10.1016/j.trip.2020.100242
18. Poliński, J., Ochociński, K.: Impact of COVID-19 on the functioning of passenger rail transport. Problemy Kolejnictwa - Railway Rep. **65**(190), 103–124 (2021). https://doi.org/10.36137/1904E
19. Rail Transport, Poland, Report on rail transport market operations 2019, Railway Transport Office, Poland (2019). https://utk.gov.pl/en/markettatistics-and-ann/market-analysis/2019/16531,Report-on-rail-transport-market-operations-2019.html
20. Almlöf, E., Rubensson, I., Cebecauer, M., Jenelius, E.: Who continued travelling by public transport during COVID-19? Socioeconomic factors explaining travel behaviour in Stockholm 2020 based on smart card data. Eur. Transp. Res. Rev. **13**(1), 31 (2021). https://doi.org/10.1186/s12544-021-00488-0
21. Beck, M.J., Hensher, D.A.: Insights into the impact of COVID-19 on household travel and activities in Australia – the early days under restrictions. Transp. Policy **96**, 76–93 (2020). https://doi.org/10.1016/j.tranpol.2020.07.001
22. Dahlberg, M., et al.: Effects of the COVID-19 pandemic on population mobility under mild policies: causal evidence from Sweden (2020). https://doi.org/10.48550/ARXIV.2004.09087
23. Rosa, G.: Travel decisions in Poland during the COVID-19 pandemic. Procedia Comput. Sci. **207**, 3103–3111 (2022). https://doi.org/10.1016/j.procs.2022.09.368
24. Szaruga, E., Pilecki, B., Sidorkiewicz, M.: The Impact of the COVID-19 pandemic, transport accessibility, and accommodation accessibility on the energy intensity of public tourist transport. Energies **16**(19), 6949 (2023). https://doi.org/10.3390/en16196949
25. University of Economics in Katowice, Katowice, Poland and A. Chwila, "The prediction of new Covid-19 cases in Poland with machine learning models. SiTns **24**(2), 59–83 (2023). https://doi.org/10.59170/stattrans-2023-020
26. Awad-Núñez, S., Julio, R., Gomez, J., Moya-Gómez, B., González, J.S.: Post-COVID-19 travel behaviour patterns: impact on the willingness to pay of users of public transport and shared mobility services in Spain. Eur. Transp. Res. Rev. **13**(1), 20 (2021). https://doi.org/10.1186/s12544-021-00476-4
27. Rail Transport, Poland, Report on rail transport market operations 2021 (2021). https://utk.gov.pl/en/markettatistics-and-ann/market-analysis/2021/19598,Report-on-rail-transport-market-operations-2021.html
28. Statistics, Poland, "Transport - activity results in 2022 (2023). https://stat.gov.pl/en/topics/transport-and-communications/transport/transport-activity-results-in-2022,6,18.html
29. Tarkhov, S.A.: Network development of the world's high-speed railways. Eur. XXI **34**, 95–112 (2018). https://doi.org/10.7163/Eu21.2018.34.6
30. Pomykała, A.: Polish regional and long-distance railway transport during the COVID-19 pandemic. Problemy Kolejnictwa **192**(2021) (2021). https://bibliotekanauki.pl/articles/2126940.pdf
31. Israel, G.D.: Determining Sample Size (1992). https://www.researchgate.net/profile/Subhash-Basu-3/post/how_could_i_determine_sample_size_for_my_study/attachment/5ebaa4924f9a520001e613b6/AS:890361492811785@1589290130539/download/samplesize1.pdf
32. Krejcie, R.V., Morgan, D.W.: Determining sample size for research activities. Educ. Psychol. Measur. **30**(3), 607–610 (1970). https://doi.org/10.1177/001316447003000308

Author Index

A
Adnan, Muhammad 106
Ahmed, Afzal 106
Ahmed, Muhammad Waqas 106
Ahmed, Muhammad 106

B
Baroch, Václav 197
Buhaiova, Maryna 55

C
Cappelli, Giuseppe 122
Celiński, Ireneusz 170

D
D'Apuzzo, Mauro 122

E
Ectors, Wim 106

F
Fikre, Girma 3

G
Gitelman, Victoria 16
Graba, Mariusz 138
Granà, Anna 89
Gruhonjić, David 30

H
Heidu, Rudolf Franz 197
Hennek, Krystian 138

J
Jakovljević, Marijan 30
Janssens, Davy 3, 106

K
Kholodov, Oleh 55
Kholodova, Olha 55
Kotylak, Piotr 210
Król, Aleksander 42
Król, Małgorzata 42

L
Leich, Andreas 75

M
Macioszek, Elżbieta 89
Mamala, Jarosław 138
Markowska, Katarzyna 225

N
Nardoianni, Sofia 122
Naumann, Sebastian 154
Neven, An 3
Nicolosi, Vittorio 122
Nippold, Ronald 75

S
Sakhno, Andrii 55
Schade, Joachim 154
Šoštarić, Marko 30
Sproch, Maciej 138
Staniek, Marcin 170
Szewczyński, Krystian 42

T
Tcheumadjeu, Louis Calvin Touko 75
Tumminello, Maria Luisa 89

V
Vertlberg, Juraj Leonard 30

W
Wagner, Peter 75
Wets, Geert 106

E. Macioszek et al. (Eds.): TSTP 2025, LNNS 1769, p. 245, 2026.
https://doi.org/10.1007/978-3-032-14094-4

MIX
Papier aus verantwortungsvollen Quellen
Paper from responsible sources
FSC® C105338

If you have any concerns about our products,
you can contact us on
ProductSafety@springernature.com

In case Publisher is established outside the EU,
the EU authorized representative is:
Springer Nature Customer Service Center GmbH
Europaplatz 3, 69115 Heidelberg, Germany

Printed by Libri Plureos GmbH
in Hamburg, Germany